Sir Henry Thompson

Food and feeding

Sir Henry Thompson

Food and feeding

ISBN/EAN: 9783337201470

Printed in Europe, USA, Canada, Australia, Japan

Cover: Foto ©berggeist007 / pixelio.de

More available books at **www.hansebooks.com**

HARPER'S HALF-HOUR SERIES.

32mo, Paper.

1. The Turks in Europe. By Edw. A. Freeman. $0 15
2, 3. Tales from Shakespeare. By Chas. and Mary Lamb. Comedies, 25 cts. Tragedies, 25 cts.
4. Thompson Hall. By Anthony Trollope. Ill's. 20
5. When the Ship Comes Home. By Walter Besant and James Rice..................... 25
6. The Life, Times, and Character of Oliver Cromwell. By E. H. Knatchbull-Hugessen.... 20
7. Early England. By F. York-Powell......... 25
8. England a Continental Power. By Louise Creighton................................ 25
9. Rise of the People, and Growth of Parliament. By James Rowley, M.A................. 25
10. The Tudors and the Reformation. By M. Creighton, M.A........................... 25
11. The Struggle Against Absolute Monarchy. By Bertha M. Cordery...................... 25
12. The Settlement of the Constitution. By Jas. Rowley, M.A............................. 25
13. England during the American and European Wars. By O. W. Tancock, M.A........... 25
14. Modern England. By Oscar Browning, M.A. 25
15. University Life in Ancient Athens. By W. W. Capes.................................. 25
16. A Primer of Greek Literature. By Eugene Lawrence............................... 25
17. A Primer of Latin Literature. By Eugene Lawrence............................... 25
18. Dieudonnée. By Geraldine Butt............ 20
19. The Time of Roses. By Geraldine Butt...... 20
20. The Jilt. By Charles Reade. Illustrated.... 20
21. The Mill of St. Herbot. By Mrs. Macquoid... 20
22. The House on the Beach. By George Meredith 20
23. Kate Cronin's Dowry. By Mrs. Cashel Hoey. 15
24. Peter the Great. By John Lothrop Motley... 25
25. Percy and the Prophet. By Wilkie Collins... 20
26. Cooking Receipts. From *Harper's Bazar*.... 25
27. Virginia. A Roman Sketch................. 25

28. The Jews and their Persecutors. By Eugene Lawrence	$0 20
29. The Sad Fortunes of the Rev. Amos Barton. By George Eliot	20
30. Mr. Gilfil's Love Story. By George Eliot	20
31. Janet's Repentance. By George Eliot	20
32. The A B C of Finance. By Simon Newcomb	25
33. A Primer of Mediæval Literature. By Eugene Lawrence	25
34. Warren Hastings. By Lord Macaulay	25
35. Addison. By Lord Macaulay	25
36. Lord Clive. By Lord Macaulay	25
37. Frederic the Great. By Lord Macaulay	25
38. The Earl of Chatham. By Lord Macaulay	25
39. William Pitt. By Lord Macaulay	25
40. Samuel Johnson, LL.D. By Lord Macaulay	25
41. Hampden.—Burleigh. By Lord Macaulay	25
42. Sir William Temple. By Lord Macaulay	25
43. Machiavelli.—Walpole. By Lord Macaulay	25
44. Milton.—Byron. By Lord Macaulay	25
45. My Lady's Money. Related by Wilkie Collins	25
46. Poor Zeph! By F. W. Robinson	20
47. Shepherds All and Maidens Fair. By Walter Besant and James Rice	25
48. Back to Back. By Edward Everett Hale	25
49. The Spanish Armada for the Invasion of England. 1587–1588. By Alfred H. Guernsey	20
50. Da Capo. By Anne Isabella Thackeray	20
51. The Bride of Landeck. By G. P. R. James	20
52. Brother Jacob.—The Lifted Veil. By Geo. Eliot	20
53. A Shadow on the Threshold. By Mary Cecil Hay	20
54. David's Little Lad. By L. T. Meade	25
55. Count Moltke's Letters from Russia	25
56. Constantinople. By James Bryce	15
57–59. English Literature Primers. By Eugene Lawrence: 57. Romance Period.—58. Classical Period.—59. Modern Period ....each	25
60. Tender Recollections of Irene Macgillicuddy	15
61. Georgie's Wooer. By Mrs. Leith-Adams	20
62. Seven Years and Mair. By Anna T. Sadlier	20
63. A Sussex Idyl. By Clementina Black	25

64. Goldsmith.—Bunyan.—Madame D'Arblay. By Lord Macaulay $0 25
65. The Youth's Health-Book 25
66. Reaping the Whirlwind. By Mary Cecil Hay. 20
67. A Year of American Travel. By Jessie Benton Frémont 25
68. A Primer of German Literature. By Helen S. Conant 25
69. The Coming Man. By Charles Reade 20
70. Hints to Women on the Care of Property 20
71. The Curate of Orsières. By Otto Roquette 20
72. The Canoe and the Flying Proa. By W. L. Alden 25
73. Back to the Old Home. By Mary Cecil Hay .. 20
74. The Lady of Launay. By Anthony Trollope. 20
75. Sir Roger de Coverley. From *The Spectator* .. 25
76. Pottery Painting. By John C. L. Sparkes 20
77. Squire Paul. By Hans Warring 25
78. Professor Pressensee. By John Esten Cooke. 25
79. The Romance of a Back Street. By F. W. Robinson 15
80. Behind Blue Glasses. By F. W. Hackländer .. 20
81. Recollections of Rufus Choate. E. P. Whipple. 15
82. Daisy Miller. By Henry James, Jr 20
83. A Primer of Spanish Literature. By Helen S. Conant 25
84. A Dark Inheritance. By Mary Cecil Hay 15
85. The Vicar of Wakefield. By Oliver Goldsmith 25
86. Stories from Virgil. By A. J. Church 25
87. Our Professor. By Mrs. E. Lynn Linton 15
88. The Sorrow of a Secret. By Mary Cecil Hay .. 15
89. Lady Carmichael's Will and other Christmas Stories. By Mary Cecil Hay and others 15
90. 'Twas in Trafalgar's Bay. By Walter Besant and James Rice 20
91. An International Episode. By H. James, Jr. 20
92. The Adventures of Ulysses. By Charles Lamb 25
93. Oliver Goldsmith's Plays 25
94. Oliver Goldsmith's Poems 20
95. Modern France. By George M. Towle 25

No.	Title	Price
96.	Our Village. By Miss Mitford	$0 25
97.	Afghanistan. By A. G. Constable	15
98.	John. By Thomas W. Knox	20
99.	The Awakening. By Mrs. Macquoid	15
100.	Ballads of Battle and Bravery	25
101.	Six Months on a Slaver. By E. Manning	20
102.	Healthy Houses. By Fleeming Jenkin. Adapted by George E. Waring, Jr	25
103.	Mr. Grantley's Idea. By John Esten Cooke	25
104.	The Four Georges. By W. M. Thackeray	25
105, 106.	The English Humorists. By W. M. Thackeray. In Two Numbers each	25
107.	Half-Hour History of England. By Mandell Creighton, M.A.	25
108.	Lord Bacon. By Lord Macaulay	25
109.	My Sister's Keeper. By Laura M. Lane	20
110.	Gaspard de Coligny. By Walter Besant, M.A.	25
111.	Tales from Euripides. By Vincent K. Cooper	20
112.	The Task. By William Cowper	20
113.	History.—Hallam's Constitutional History. By Lord Macaulay	25
114.	The Lay of the Last Minstrel. By Sir Walter Scott	20
115.	Marmion. By Sir Walter Scott	25
116.	The Lady of the Lake. By Sir Walter Scott	25
117.	The Lover's Tale. By Alfred Tennyson	10
118.	Wassail. By Colonel Charles Hamley	20
119.	Modern Whist. By Fisher Ames	20
120.	The Rivals and the School for Scandal. By Richard Brinsley Sheridan	25
121.	Holidays in Eastern France. By M. Betham-Edwards	25
122.	Labor and Capital Allies—Not Enemies. By Edward Atkinson	20
123.	Chapters on Ants. By Mary Treat	20
124.	The Bar-Maid at Battleton. By F. W. Robinson	15

FOOD AND FEEDING

BY

SIR HENRY THOMPSON

NEW YORK
HARPER & BROTHERS, PUBLISHERS
FRANKLIN SQUARE
1879

FOOD AND FEEDING.

I THINK I shall not be far wrong if I say that there are few subjects more important to the well-being of man than the selection and preparation of his food. Our forefathers in their wisdom have provided, by ample and generously endowed organizations, for the dissemination of moral precepts in relation to human conduct, and for the constant supply of sustenance to meet the cravings of religious emotions common to all sorts and conditions of men. In these provisions no student of human nature can fail to recognize the spirit of wisdom and a lofty purpose. But it is not a sign of ancestral wisdom that so little thought has been bestowed on the teaching of what we should eat and drink; that the relations, not only between food and a healthy population but between food and virtue, between the process of digestion and the state of mind which results from it, have occupied a subordinate place in the practical arrangements of life. No doubt there has long been some practical acknowledgment, on the part of a few educated persons, of the simple fact that a man's temper, and consequently many of his actions, depend on such an alternative as whether he habitually digests his food well or ill; whether the meals which he eats are properly converted into healthy material, suitable for the ceaseless work of building up both muscle and brain; or whether unhealthy products constantly pollute the course of nutritive supply. But the truth of that fact has never been generally admitted to an extent at all comparable with its exceeding importance. It produces no practical result on the habits of men in the least degree commensurate with the pregnant import it contains. For it is certain that an adequate recognition of the value of proper food to the individual in maintaining a high standard of health, in prolonging healthy life (the prolongation of unhealthy life being small gain either to the individual or to the community), and thus largely promoting cheerful temper, prevalent good-nature, and improved moral tone, would require almost a revolution in the habits of a large part of the community.

The general outlines of a man's mental character and physical tendencies are, doubtless, largely determined by the impress of race and family. That is, the scheme of the building, its characteristics and dimensions, are inherited; but, to a very large extent, the materials and filling in of the framework depend upon his food and training. By the latter term may be understood all that relates to mental and moral, and even to physical, education, in part already assumed to be fairly provided for, and therefore not further to be considered here. No matter, then, how consummate the scheme of the architect, or how vast the design, more or less of failure to rear the edifice results when the materials are ill chosen, or wholly unworthy to be used. Many other sources of failure there may be which it is no part of my business to note; but the influence of food is not only itself cardinal in rank, but, by priority of action, gives rise to other and secondary agencies.

The slightest sketch of the commonest types of human life will suffice to illustrate this truth.

To commence, I fear it must be admitted that the majority of British infants are reared on imperfect milk, by weak or ill-fed mothers; and thus it follows that the signs of disease, of feeble vitality, or of fretful disposition, may be observed at a very early age, and are apparent in symptoms of indigestion, or in the cravings of want manifested by the "peevish" and sleepless infant. In circumstances where there is no want of abundant nutriment, overfeeding or complicated forms of food, suitable only for older persons, produce for this infant troubles which are no less grave than those of the former. In the next stage of life, among the poor, the child takes his place at the parents' table, where lack of means, as well as of knowledge, deprives him of food more suitable than the rough fare of the adult, and, moreover, obtains for him, perchance, his little share of beer or gin. On the whole, perhaps he is not much worse off than the child of the well-to-do, who becomes a pet, and is already familiarized with complex and too solid forms of food and stimulating drinks, which custom and self-indulgence have placed on the daily table. And soon afterward commence in consequence—and entirely in consequence, a fact it is impossible too much to emphasize—the "sick headaches" and "bilious attacks" which pursue their victim through half a lifetime, to be exchanged for gout, or worse, at or before the grand climacteric. And so common are these evils that they are regarded by people in general as a necessary appanage of "poor humanity." No notion can be more erroneous, since it is absolutely true that the complaints referred to are self-engendered, form no necessary part of our physical nature, and for their existence are dependent almost entirely on our habits in relation to food and drink. I except, of course, those cases in which hereditary tendencies are so strong as to produce these evils, despite of some care on the part of the unfortunate victim of an ancestor's self-indulgence. Equally, however, on the part of that little-to-be-revered progenitor was ill-chosen food, or, more probably, excess in quantity, the cause of disease, and not the physical nature of man.

The next stage of boyhood transfers the child just spoken of to a public school, where too often insufficient or inappropriate diet, at the most critical period of growth, has to be supplemented from other sources. It is almost unnecessary to say that chief among these are the pastry-cook and the vender of portable provisions, for much of which latter that skin-stuffed compound of unknown origin, an uncertified sausage, may be accepted as the type.

After this period arise the temptations to drink, among the youth of all classes, whether at beer-house, tavern, or club. For it has been taught in the bosom of the family, by the father's example and by the mother's precept, that wine, beer, and spirits are useful, nay, necessary, to health, and that they augment the strength. And the lessons thus inculcated and too well learned were but steps which led to wider experience in the pursuit of health and strength by larger use of the same means. Under such circumstances it often happens, as the youth grows up, that a flagging appetite or a failing digestion habitually demands a dram before or between meals, and that these are regarded rather as occasions to indulge in variety of liquor than as repasts for nourishing the body. It is not surprising, with such training, that the true object of both eating and drinking is entirely lost sight of. The gratification of acquired tastes usurps the function of that zest which healthy appetite produces; and the intention that food should be adapted to the physical needs of the body and the healthy action of the mind is forgotten altogether. So it often comes to pass that at middle-age, when man finds himself in the full current of life's occupations, struggling for pre-eminence with his fellows, indigestion has become persistent in some of its numerous forms, shortens his "staying power," or spoils his judgment or temper. And, besides all this, few causes are more potent than an incompetent stomach to engender habits of selfishness and egotism. A constant care to provide little personal wants of various kinds, thus rendered necessary, cultivates these sentiments, and they influence the man's whole character in consequence. The poor man, advancing in years, suffers from continuous toil with inadequate food, the supply of which is often diminished by his expenditure for beer, which, although often noxious, he regards as the elixir of life, never to be missed when fair occasion for obtaining it is offered. Many of this class are prematurely crippled by articular disease, etc., and become permanent inmates of the parish workhouse or infirmary.

It must be obvious to everybody how much more of detail might be added to fill in the outlines of this little sketch. It is meagre in the extreme; nevertheless it suffices for my purpose: other illustrations may occur hereafter.

But it is necessary to say at this point, and I desire to say it emphatically, that the subject of food need not, even with the views just enunciated, be treated in an ascetic spirit. It is to be considered in relation to a principle, in which we may certainly believe, that aliments most adapted to develop the individual, sound in body and mind, shall not only be most acceptable, but that they may be selected and prepared so as to afford scope for the exercise of a refined taste, and produce a fair degree of that pleasure naturally associated with the function of the palate, and derived from a study of the table. For it is certain that nine-tenths of the gormandize which is practised, at all events in English society — where, for the most part, it is a matter of faith without knowledge—is no more a source of gratification to the eater's gustatory sense than it is of digestible sustenance to his body.

Our subject now shapes itself. Food must first be regarded in relation to its value as material to be used for building up and sustaining that composite structure, the human body, under the varied conditions in which it may be placed. Secondly, the selection of food, and the best modes of preparing it, resulting in the production of "the dish," a subject of great extent and importance, must be dealt with very briefly. Lastly, the exercise of taste in relation to the serving of food and drink, or the art of combining dishes to form " a meal," must also be considered.

I shall not regard this as the place in which to offer any scientific definition of the term food. I shall include within its range all the solid materials popularly so regarded and therefore eaten. And drink being as necessary as solids for the purpose of digestion, and to supply that large proportion of fluid which the body contains in every mesh and cell thereof, I shall regard as "drink" all the liquids which it is customary to swallow with our meals, although probably very few, if any, of them can be regarded as food, in any strict sense of the term.

Food is essential to the body in order to fulfil two distinct purposes, or to supply two distinct wants inseparable from animal life. As certainly as a steam-engine requires fuel, by the combustion or oxidation of which force is called into action for various purposes—as the engine itself requires the mending and replacing of parts wasted in the process of working—so certainly does the animal body require fuel to evoke its force, and material to replace those portions which are necessarily wasted by labor, whether the latter be what we call physical or mental, that is, of limbs or of brain. The material which is competent to supply both requirements is a complete or perfect food. Examples of complete food exist in milk and the egg, sufficing as these do for all the wants of the young animal during the period of early growth. Nevertheless a single animal product like either of the two named, although complex in itself, is not more perfect than an artificial combination of various simpler substances, provided the mixture (dish or meal) contains all the elements required in due proportion for the purposes of the body.

It would be out of place to occupy much

space with those elementary details of the chemical constitution of the body which may be found in any small manual of human physiology;[1] but, for the right understanding of our subject, a brief sketch must be presented. Let it suffice to say that carbon, hydrogen, and oxygen, the three all-pervading elements of the vegetable world, enter largely into the composition of the animal body; and that the two former especially constitute a fuel, the oxidation of which produces animal heat, and develops the force in its varied forms, physical and mental, which the body is capable of exerting. Besides these, nitrogen, obtainable from certain vegetable products, not from all, but forming definite combinations with the three elements just named, is essential to the repair and reproduction of the body itself, being one of its most important con-

---

[1] Such as *Physiology*, Science Primer, by M. Foster, M.A., M.D. (Macmillan); *Lessons in Elementary Physiology*, by Professor Huxley (Macmillan). For a full consideration of the subject, Dr. Pavy's very complete *Treatise on Food and Dietetics* (Churchill, London, 1875).

stituents. Lastly must be named several other elements, which, in small proportions, are also essential constituents of the body; such as sulphur, phosphorus, salts of lime, magnesia, potash, etc., with traces of iron and other metals. All these must be present in the food supplied, otherwise animal existence cannot be supported; and all are found in the vegetable kingdom, and may be obtained directly therefrom by man in feeding on vegetables alone.[1] But the process of obtaining and combining these simple elements into the more complex forms which constitute the bases of animal tissues—definite compounds of nitrogen with carbon, hydrogen, and oxygen—is also accomplished by the lower animals, which are exclusively vegetable feeders. These animals we can consume as food, and thus procure, if we please, ready prepared for our purpose,

---

[1] The vegetable kingdom comprehends the cereals, legumes, roots, starches, sugar, herbs, and fruits. Persons who style themselves vegetarians often consume milk, eggs, butter, and lard, which are choice foods from the animal kingdom. There are other persons, of course, who are strictly vegetable eaters, and such alone have any right to the title of vegetarians.

the materials of flesh, sinew, and bone, for immediate use. We obtain also from the animal milk and the egg, already said to be "perfect" foods; and they are so because they contain the nitrogenous compounds referred to, fatty matter abundantly, and the various saline or mineral matters requisite. But compounds simpler in form than the preceding, of a non-nitrogenous kind, that is, of carbon, hydrogen, and oxygen only, are necessary as food for the production of animal heat and force. These consist, first, of the fat of animals of various kinds, and of butter; and, from the vegetable kingdom, of the fatty matter which exists in grain and legumes, and largely in the olive and in many seeds: secondly, of the starchy matters, all derived from vegetables, such as a large part of wheaten and other flour, rice, arrowroot, and potatoes; together with sugar, gum, and other minor vegetable products of a similar kind. The fats form the more important group of the two, both in relation to the production of heat and force; and without a constant supply of fat as food the body would cease to exist. The vegetable eater, pure and simple, can therefore extract from his food all the principles necessary for the growth and support of the body as well as for the production of heat and force, provided that he selects vegetables which contain all the essential elements named. But he must for this purpose consume the best cereals, wheat or oats; or the legumes, beans, pease, or lentils; or he must swallow and digest a large weight of vegetable matter of less nutritive value, and therefore containing at least one element in large excess, in order to obtain all the elements he needs. Thus the Irishman requires ten to eleven pounds of potatoes daily, which contain chiefly starch, very little nitrogen, and scarcely any fat: hence he obtains, when he can, some buttermilk, or bacon, or a herring, to supply the deficiency. The Highlander, living mainly on oatmeal, requires a much smaller weight, since this grain contains not only starch but much nitrogen and a fair amount of fat, although not quite sufficient for his purpose, which is usually supplied by adding milk or a little bacon to his diet.

On the other hand, the man who lives chiefly or largely on flesh and eggs, as well as bread, obtains precisely the same principles, but served in a concentrated form, and a weight of about two or three pounds of such food is a full equivalent to the Irishman's ten or eleven pounds of potatoes and extras. The meat eater's digestion is taxed with a far less quantity of solid, but that very concentration in regard of quality entails on some stomachs an expenditure of force in digestion equal to that required by the vegetable eater to assimilate his much larger portions. And it must be admitted, as a fact beyond question, that some persons are stronger and more healthy who live chiefly or altogether on vegetables, while there are many others for whom a proportion of animal food appears to be desirable, if not necessary. In studying this matter, individual habit must be taken into account. An animal feeder may by slow degrees become a vegetarian, without loss of weight or strength, not without feeling some inconvenience in the process; but a sudden change in diet in this direction is for a time almost equivalent to starvation. The digestive organs require a considerable period to accommodate themselves to the performance of work different from that to which they have been long accustomed, and in some constitutions might fail altogether in the attempt. Besides, in matters of diet essentially, many persons have individual peculiarities; and while certain fixed principles exist, such as those already laid down as absolutely cardinal, in the detail of their application to each man's wants, an infinity of stomach-eccentricities is to be reckoned on. The old proverb expresses the fact strongly but truly, "What is one man's meat is another man's poison." Yet nothing is more common—and one rarely leaves a social dinner-table without observing it—than to hear some good-natured person recommending to his neighbor, with a confidence rarely found except in alliance with profound ignorance of the matter in hand, some special form of food, or drink, or system of diet, solely because the adviser happens to have found it useful to himself.

It will be interesting now to take a general but brief survey of the vast range of materials which civilized man has at his command for the purpose of food: these few preliminary remarks on the chemical constituents of food having been intended to aid in appreciating the value of different kinds.

Commencing with the vegetable kingdom, from which our early progenitors, probably during long ages, drew all their sustenance, the cereals, or cultivated grasses, come first, as containing all the elements necessary to life, and being, therefore, most largely consumed. Wheat and its congeners, which rank highest in quality, had been distinguished, in the form of bread, as "the staff of life," long before the physiological demonstration of the fact had been attained. Wheat, oats, rye and barley, maize and rice, are the chief members of this group; wheat containing the most nitrogenous or flesh-forming material, besides abundance of starch, a moderate amount of fat, together with sufficient saline and mineral elements. Rice, on the other hand, contains very little nitrogen, fat, and mineral constituents, but starch in great abundance; while maize, with a fair supply of nitrogenous and starchy matter, contains the most fat or heat-producing material of the whole group. As derived from wheat must be named the valuable aliments, macaroni and all the Italian pastes. Derived from barley is malt-saccharine, parent of the large family of fermented liquors known as beer. And from various other grains are obtained, by fermentation and distillation, several forms of ardent spirit. Vinegar, best when produced from the grape, is also largely made from grain.

The legumes, such as beans, lentils, and pease, form an aliment of great value, containing more nitrogen even than the cereals, but with fat in very small proportion, while starchy matter and the mineral elements abound in both groups.

The tuber finds its type in the potato, which contains much starch, little nitrogen, and almost no fat; in the yam also. The roots may be illustrated by the beet, carrot, parsnip, and turnip, all containing little nitrogen, but much sugar, and water in large proportion. Derived from roots and stems of foreign growth, we have arrowroot, tapioca, and sago, all starches, and destitute of nitrogen. Fatty matter is abundantly found in the olive, which supplies a large part of the world with an important article of food. The almond and other seeds are also fruitful sources of oil.

Under the term "green vegetables," a few leading plants may be enumerated as types of the vast natural supplies which everywhere exist: The entire cabbage tribe in great variety; lettuces, endive, and cresses; spinach, sea-kale, asparagus, celery, onions, artichokes, and tomato, all valuable, not so much for nutritive property, which is inconsiderable, as for admixture with other food, chiefly on account of salts which they contain, and for their appetizing aroma and varied flavors. Thus condiments are useful, as the sweet and aromatic spices, the peppers, mustard, and the various potherbs, so essential to an agreeable cuisine. Sea-weeds, as laver, and the whole tribe of mushrooms, should be named, as ranking much higher in nutritive value than green vegetables. Pumpkins, marrows, and cucumbers, chestnuts and other nuts, largely support life in some countries. The bread-fruit is of high value; so also are the cocoa-nut and the banana in tropical climates.

Lastly must be named all those delicious but not very nutritive products of most varied kind and source, grouped under the name of fruits. These are characterized chiefly by the presence of sugar, acid, vegetable jelly, and some saline matter, often combined with scent and flavor of exquisite quality. Derived from grapes as its chief source, stands wine in its innumerable varieties, so closely associated by all civilized nations with the use of aliments, although not universally admitted to rank in technical language as a food. Next may be named sugar in its various forms, a non-nitrogenous product of great value, and, in a less degree, honey. No less important are the tea-plant, the coffee-berry, and the seeds of the cacao-tree.

There is a single element belonging to the mineral kingdom which is taken in its natural state as an addition to food, namely, common salt; and it is so universally recognized as necessary, that it cannot be omitted here. The foregoing list possesses no claim to be exhaustive, only to be fairly typical and suggestive; many omissions, which some may think important, doubtless exist. In like manner, a rapid survey may be taken of the animal kingdom.

First, the flesh of domestic quadrupeds: the ox and sheep, both adult and young; the pig; the horse and ass, chiefly in France. Milk, butter, and cheese in endless variety are derived chiefly from this group. More or less wild are the red deer, the fallow-deer, and the roe-deer. As game, the hare and rabbit; abroad, the bison, wild-boar, bear, chamois, and kangaroo, are esteemed for food among civilized nations; but many other animals are eaten by half-civilized and savage peoples. All these are rich in nitrogen, fatty matters, and saline materials.

Among birds, we have domestic poultry in great variety of size and quality, with eggs in great abundance furnished chiefly

FOOD AND FEEDING.

by this class. All the wild-fowl and aquatic birds; the pigeon tribe and the small birds. Winged game in all its well-known variety.

Of fish it is unnecessary to enumerate the enormous supply and the various species which exist everywhere, and especially on our own shores, from the sturgeon to whitebait, besides those in fresh-water rivers and lakes. All of them furnish nitrogenous matter largely, but, and particularly the white-fish, possess fat in very small proportion, and little of saline materials. The salmon, mackerel, and herring tribes have more fat, the last named in considerable quantity, forming a useful food well calculated to supplement cereal aliments, and largely adopted for the purpose both in the south and north of Europe.

The so-called reptiles furnish turtle, tortoise, and edible frog. Among articulated animals are the lobsters, crabs, and shrimps. Among mollusks, the oyster and all the shell-fish, which, as well as the preceding animals, in chemical composition closely resemble that of fish properly so called.

I shall not enter on a discussion of the question: Is man designed to be a vegetable feeder, or a flesh-eating or an omnivorous animal? Nor shall his teeth or other organs be examined in reference thereto. Any evidence to be found by anatomical investigation can only be safely regarded as showing what man is and has been. That he has been and is omnivorous to the extent of his means, there can exist no doubt. Whether he has been generally prudent or happy in his choice of food and drink is highly improbable, seeing that until very recently he has possessed no certain knowledge touching the relations which matters used as food hold with respect to the structure and wants of his body, and that such recent knowledge has been confined to a very few individuals. Whatever sound practice he may have attained, and it is not inconsiderable, in his choice and treatment of food, is the result of many centuries of empirical observation, the process of which has been attended with much disastrous failure and some damage to the experimenters. No doubt much unsound constitution and proclivity to certain diseases result from the persistent use through many generations of improper food and drink.

Speaking in general terms, man seems, at the present time, prone to choose foods which are unnecessarily concentrated and too rich in nitrogenous or flesh-forming material, and to consume more in quantity than is necessary for the healthy performance of the animal functions. He is apt to leave out of sight the great difference, in relation to both quantity and quality of food, which different habits of life demand, *e. g.*, between the habits of those who are chiefly sedentary and brain-workers, and of those who are active, and exercise muscle more than brain. He makes very small account of the different requirements by the child, the mature adult, and the declining or aged person; and he seems to be still less aware of the frequent existence of notable individual peculiarities in relation to the tolerance of certain aliments and drinks. As a rule, man has little knowledge of, or interest in, the processes by which food is prepared for the table, or the conditions necessary to the healthy digestion of it by himself. Until a tolerably high standard of civilization is reached, he cares more for quantity than quality, desires little variety, and regards as impertinent an innovation in the shape of a new aliment, expecting the same food at the same hour daily, his enjoyment of which apparently greatly depends on his ability to swallow the portion with extreme rapidity, that he may apply himself to some other and more important occupation without delay. Eating is treated, in fact, by multitudes much as they are disposed to treat religious duty—which eating assuredly is—that is, as a duty which is generally irksome, but unfortunately necessary to be performed. As to any exercise of taste in the serving or in the combining of different foods at a meal, the subject is completely out of reach of the great majority of people, and is as little comprehended by them as the structure and harmonies of a symphony are by the first whistling boy one chances to meet in the street. The intelligent reader who has sufficient interest in this subject to have followed me thus far may fancy this a sketch from savage life. On the contrary, I can assure him that ignorance and indifference to the nature and object of food mark the condition of a large majority of the so-called educated people of this country. Men even boast of their ignorance of so trivial a subject, regard it as unworthy the exercise of their powers, and—a small compliment to their wives and sisters—fit only for the occupation of women.

Admitting man, then, to be physically so constituted as to be able to derive all that is necessary to the healthy performance of all his functions from the animal or from the vegetable kingdom, either singly or combined, he can scarcely be regarded otherwise than as qualified to be an omnivorous animal. Add to this fact his possession of an intelligence which enables him to obtain food of all kinds and climes, to investigate its qualities, and to render it more fit for digestion by heat — powers which no other animal possesses—and there appears no *à priori* reason for limiting his diet to products of either kingdom exclusively.

It is a matter of great interest to ascertain what have become, under the empirical conditions named, the staple foods of the common people of various climates and races—what, in short, supports the life and labor of the chief part of the world's population.

In the tropics and adjacent portions of the temperate zones, high temperature being incompatible with the physical activity familiar to Northern races, a very little nitrogenous material suffices, since the waste of muscle is small. Only a moderate quantity of fat is taken, the demand for heat-production being inconsiderable. The chiefly starchy products supply nearly all the nutriment required, and such are found in rice, millet, etc. Rice by itself is the principal food of the wide zone thus indicated, including a large part of China, the East Indies, part of Africa and America, and also the West Indies. Small additions, where obtainable, are made of other seeds, of oil, butter, etc.; and as temperature decreases by distance from the equator, some fish, fowl, or other light form of animal food, are added.

In the north of Africa, Arabia, and some neighboring parts, the date, which contains sugar in abundance, is largely eaten, as well as maize and other cereals.

Crossing to Europe, the southern Italian is found subsisting on macaroni, legumes, rice, fruits, and salads, with oil, cheese, fish, and small birds, but very little meat. More northward, besides fish and a little meat, maize is the chief aliment, rye and other cereals taking a second place. The chestnut, also, is largely eaten by the poorer population, both it and maize containing more fatty matter than wheat, oats, and legumes.

In Spain, the inhabitants subsist chiefly on maize and rice, with some wheat and legumes, among them the garbanzo or "chickpea," and one of the principal vegetable components of the national *olla*, which contains also a considerable proportion of animal food in variety, as bacon, sausage, fowl, etc. Fruit is fine and abundant; especially so are grapes, figs, and melons. There is little or no butter, the universal substitute for which is olive oil, produced in great quantity. Fowls and the pig furnish the chief animal food, and garlic is the favorite condiment.

Going northward, flesh of all kinds occupies a more considerable place in the dietary. In France the garden vegetables and legumes form an important staple of diet for all classes; but the very numerous small land proprietors subsist largely on the direct products of the soil, adding little more than milk, poultry, and eggs, the produce of their small farms. The national *pot-au-feu* is an admirable mixed dish, in which a small portion of meat is made to yield all its nutritive qualities, and to go far in mingling its odor and savor with those of the fragrant vegetables which are so largely added to the stock. The stock-meat eaten hot, or often cold with plenty of green salad and oil, doubtless the most palatable mode of serving it, thus affords a source of fat, if not otherwise provided for by butter, cheese, etc.

Throughout the German Empire, the cereals, legumes, greens, roots, and fruits supply an important proportion of the food consumed by the common population. Wheaten bread chiefly, and some made from rye, also beans and pease, are used abundantly. Potatoes and green vegetables of all kinds are served in numerous ways, but largely in soups, a favorite dish. Meats, chiefly pork, are greatly esteemed in the form of sausage, and appear also as small portions or joints, but freely garnished with vegetables, on the tables of those who can afford animal diet. Going northward, where the climate is no longer adapted for the production of wheat, as in parts of Russia, rye and oats form the staple food from the vegetable kingdom, associated with an increased quantity of meat and fatty materials.

Lastly, it is well known that the inhabitants of the arctic zone are compelled to consume large quantities of oily matter, in order to generate heat abundantly; and also that animal food is necessarily the staple of their dietary. Vegetables, which moreover are not producible in so severe a climate, would there be wholly inadequate to support life.

We will now consider the food which the English peasant and artisan provide. The former lives, for the most part, on wheaten bread and cheese, with occasionally a little bacon, some potatoes, and perhaps garden greens: it is rarely, indeed, that he can obtain flesh. To this dietary the artisan adds meat, mostly beef or mutton, and some butter. A piece of fresh and therefore not tender beef is baked, or cooked in a frying-pan, in the latter case becoming a hard, indigestible, and wasted morsel; by the former process a somewhat better dish is produced, the meat being usually surrounded by potatoes or by a layer of some batter, since both contain starchy products, and absorb the fat which leaves the meat. The food of the peasant might, however, be cheaper and better; while the provision of the artisan is simply extravagant and bad. At this period of our national history, when food is scarce, and the supply of meat insufficient to meet the demand which our national habits of feeding perpetuate, it is an object of the first importance to consider whether other aliments can be obtained at a cheaper rate, and at the same time equal in quality to those of the existing dietary. Many believe that this object may be accomplished without difficulty, and that the chief obstacle to improvement in the food-supply, not only of the classes referred to, but in that of the English table generally, is the common prejudice which exists against any aliment not yet widely known or tried. The one idea which the working-classes possess in relation to improvement in diet, and which they invariably realize when wages are high, is the inordinate use of butcher's meat. To make this the chief element of at least three meals daily, and to despise bread and vegetables, is for them no less a sign of taste than a declaration of belief in the perfection of such food for the purposes of nutrition.

We have already seen that not only can all that is necessary to the human body be supplied by the vegetable kingdom solely, but that, as a matter of fact, the world's population is to a large extent supported by vegetable products. Such form, at all events, the most wholesome and agreeable diet for the inhabitants of the tropics. Between about forty and nearly sixty degrees of latitude we find large populations of fine races trained to be the best laborers in the world on little more than cereals and legumes, with milk (cheese and butter), as food; that little consisting of irregular and scanty supplies of fish, flesh, and fatty matter. In colder regions vegetable products are hardly to be obtained, and flesh and fat are indispensable. Thus man is clearly omnivorous; while *men* may be advantageously vegetarian in one climate, mixed eaters in another, and exclusively flesh-eaters in a third.

I have not hesitated to say that Englishmen generally have adopted a diet adapted for a somewhat more northerly latitude than that which they occupy; that the cost of their food is therefore far greater than it need be, and that much of their peculiar forms of indigestion and resulting chronic disease is another necessary consequence of the same error. They consume too much animal food, particularly the flesh of cattle. For all who are occupied with severe and continuous mechanical labor, a mixed diet, of which cereals and legumes form a large portion, and meat, fish, eggs, and milk form a moderate proportion, is more nutritious and wholesome than chiefly animal food. For those whose labor is chiefly mental, and whose muscular exercise is inconsiderable, still less of concentrated nitrogenous food is desirable. A liberal supply of cereals and legumes, with fish, and flesh in its lighter forms, will better sustain such activity than large portions of butcher's-meat twice or thrice a day. Then, again, it is absolutely certain, contrary to the popular belief as this is, that while a good supply of food is essential during the period of growth and active middle life, a diminished supply is no less essential to health and prolongation of life during declining years, when physical exertion is small, and the digestive faculty sometimes becomes less powerful also. I shall not regard it as within my province here to dilate on this topic, but shall assert that the "supporting" of aged persons, as it is termed, with increased quantities of food and stimulant, is an error of cardinal importance. These things being so, a consideration of no small concern arises in relation to the economical management of the national resources. For it is a fair computation that every acre of land devoted to the production of meat is capable of becoming the source of three or four times the amount of produce of equivalent value as food, if devoted to the production of grain. In other words, a given area of land cropped with cereals and legumes will support a population more than three times as numerous as that which can be sustained on the same land devoted to the growth of cattle. Moreover, the cornland will produce, almost without extra cost, a considerable quantity of animal food, in the form of pigs and poultry, from the offal or coarser parts of vegetable produce which is unsuitable for human consumption.

Thus this country purchases every year a large and increasing quantity of corn and flour from foreign countries, while more of our own land is yearly devoted to grazing purposes. The value of corn and flour imported by Great Britain in 1877 was no less than £63,536,322, while in 1875 it was only just over £53,000,000. The increased import during the last thirty-two years is well exhibited in the following extract: "In 1846 the imports of corn and flour amounted to seventeen pounds' weight per head of population; in 1855 they had risen to seventy pounds per head; and in 1865 to ninety-three pounds' weight per head of population. Finally, in 1877 the imports of corn and flour amounted to one hundred and seventy pounds' weight per head of population of the United Kingdom."[1]

Lastly, those who are interested in the national supply of food must lament that, while Great Britain possesses perhaps the best opportunities in the world for securing a large and cheap supply of fish, she fails to attain it, and procures so little only, that it is to the great majority of the inhab-

---

[1] Statesman's Year-Book, 1879, p. 258.

itants an expensive luxury. Fish is a food of great value; nevertheless, it ought, in this country, to be one of the cheapest aliments, since production and growth cost absolutely nothing, only the expenses of catching and of a short transport being incurred.

Having enunciated some general principles which it is important should first be established, I shall offer briefly an illustration or two of the manner in which they may be applied. This brings us to the second division of the subject, viz., the practical treatment of certain aliments in order to render them suitable for food. Dealing first with that of the agricultural laborer, our object is to economize his small pittance, to give him, if possible, a rather more nutritive, wholesome, and agreeable dish— he can have but one—than his means have hitherto furnished. But here there is little scope for change; already said to live chiefly on bread and cheese, with occasionally bacon, two indications only for improvement can be followed, viz., augmentation of nitrogenous matter to meet the wear and tear of the body in daily hard labor, and of fatty matter to furnish heat and force. A fair proportion of meat, one of the best means of fulfilling them, is not within his reach. First, his daily bread ought to contain all the constituents of the wheat, instead of being made of flour from which most of the mineral elements have been removed. But beans and pease are richer in nitrogen than wheat, and equal it in starch, mineral matters, and fat, the last being in very small quantity, while maize has three times their proportion of fat. Hence all of these would be useful additions to his dietary, being cheaper than wheat in the market, although, the retail demand being at present small, they may not be so in the small shops. As an illustration of the value of legumes combined with fat, it may be remembered how well the erbswurst supported the work of the German armies during the winter of 1870–'71, an instructive lesson for us in England at the present moment. It consists of a simple pea-soup mixed with a certain proportion of bacon or lard, and dried so as to be portable, constituting in very small compass a perfect food, especially suitable for supporting muscular expenditure and exposure to cold. Better than any flesh, certainly any which could be transported with ease, the cost was not more than half that of ordinary meat. It was better also, because the form of the food is one in which it is readily accessible and easily digested. It was relished cold, or could be converted in a few minutes into good soup with boiling water. But for our laborer probably the best of the legumes is the haricot bean, red or white, the dried mature bean of the plant whose pods we eat in the early green state as "French beans." For this purpose they may be treated thus: Soak, say, a quart of the dried haricots in cold water for about twelve hours, after which place them in a saucepan, with two quarts of cold water and a little salt, on the fire; when boiling, remove to the corner and simmer slowly until the beans are tender, the time required being about two to three hours. This quantity will fill a large dish, and may be eaten with salt and pepper. It will be greatly improved at small cost by the addition of a bit of butter, or of melted butter with parsley, or if an onion or two have been sliced and stewed with the haricots. A better dish still may be made by putting all or part, after boiling, into a shallow frying-pan, and lightly frying for a few minutes with a little lard and some sliced onions. With a few slices of bacon added, a comparatively luxurious and highly nutritive meal may be made. But there is still in the saucepan, after boiling, a residue of value, which the French peasant's wife, who turns everything to account, utilizes in a manner quite incomprehensible to the Englishwoman. The water in which dried haricots have stewed, and also that in which green French beans have been boiled, contain a proportion of nutritious matter. The Frenchwoman always preserves this liquor carefully, cuts and fries some onions, adds these and some thick slices of bread, a little salt and pepper, with a pot-herb or two from the corner of the garden, and thus serves hot an agreeable and useful *croûte au pot*. It ought to be added that the haricots so largely used by the working-classes throughout Europe are not precisely either "red" or "white," but some cheaper local varieties, known as *haricots du pays*. These, I am assured on good authority, could be supplied here at about twopence a pound, their quality as food being not inferior to other kinds.

But haricots—let them be the fine white Soissons—are good enough to be welcome at any table. A roast leg or shoulder of mutton should be garnished by a pint boiled as just directed, lying in the gravy of the dish; and some persons think that, with a good supply of the meat gravy, and a little salt and pepper, "the haricots are by no means the worst part of the mutton." Then with a smooth *purée* of mild onions, which have been previously sliced, fried brown, and stewed, served freely as sauce, our leg of mutton and haricots become the *gigot à la bretonne* well known to all lovers of wholesome and savory cookery. Next, white haricots stewed until soft, made into a rather thick *purée*, delicately flavored by adding a small portion of white *purée* of onions (not browned by frying as in the preceding sauce), produce an admirable garnish for the centre of a dish of small cutlets, or an *entrée* of fowl, etc. Again, the same haricot *purée* blended with a veal stock, well flavored with fresh vegetables, furnishes an admirable and nutritious white soup. The red haricots in like manner, with a beef stock, make a superlative brown soup, which, with the addition of portions of game, especially of hare, forms, for those who do not despise economy in cuisine where the result attained is excellent, a soup which in texture and in flavor would by many persons not be distinguishable from a common *purée* of game itself. Stewed haricots also furnish, when cold, an admirable salad, improved by adding slices of tomato, etc., the oil supplying the one element in which the bean is deficient; and a perfectly nutritious food is produced for those who can digest it—and they are numerous—in this form. The same principle, it may be observed, has, although empirically, produced the well-known dishes of beans and bacon, ham and green pease, boiled pork and pease-pudding, all of them old and popular, but scientific, combinations. Thus also the French, serving *petits pois* as a separate dish, add butter freely and a dash of sugar, the former making the compound physiologically complete, the latter agreeably heightening the natural sweetness of the vegetable.

Let me recall, at the close of these few hints about the haricot, the fact that there is no product of the vegetable kingdom so nutritious; holding its own in this respect, as it well can, even against the beef and mutton of the animal kingdom. The haricot ranks just above lentils, which have been so much praised of late, and rightly, the haricot being also to most palates more agreeable. By most stomachs, too, haricots are more easily digested than meat is; and, consuming weight for weight, the eater feels lighter and less oppressed, as a rule, after the leguminous dish; while the comparative cost is very greatly in favor of the latter. I do not, of course, overlook in the dish of simple haricots the absence of savory odors proper to well-cooked meat; but nothing is easier than to combine one part of meat with two parts of haricots, adding vegetables and garden herbs, so as to produce a stew which shall be more nutritious, wholesome, and palatable than a stew of all meat with vegetables, and no haricots. Moreover, the cost of the latter will be more than double that of the former.

I have just adverted to the bread of the laborer, and recommended that it should be made from entire wheat meal; but it should not be so coarsely ground as that commonly sold in London as "whole meal." The coarseness of "whole meal" is a condition designed to exert a specific effect on the digestion for those who need it, and, useful as it is in its place, is not desirable for the average population referred to. It is worth observing, in relation to this coarse meal, that it will not produce light, agreeable bread in the form of loaves: they usually have either hard flinty crusts, or soft dough-like interiors; but the following treatment, after a trial or two, will be found to produce excellent and most palatable bread: To two pounds of whole meal add half a pound of fine flour and a sufficient quantity of baking-powder and salt; when these are well mixed, rub in about two ounces of butter, and make into dough with half milk and water, or with all milk if preferred. Make rapidly into flat cakes, like "tea-cakes," and bake in a quick oven, leaving them afterward to finish thoroughly at a lower temperature. The butter and milk supply fatty matter, in which the wheat is somewhat deficient; all the saline and mineral matters of the husk are retained; and thus a more nutritive form of bread cannot be made. Moreover, it retains the natural flavor of the wheat, in place of the insipidity which is characteristic of fine flour, although it is indisputable that bread produced from the latter, especially at Paris and Vienna, is unrivalled for delicacy, texture, and color. Whole meal may be bought; but mills are now cheaply made for home use, and wheat may be ground to any degree of coarseness desired.

Here illustration by recipe must cease;

although it would be an easy task to fill a volume with matter of this kind, illustrating the ample means which exist for diminishing somewhat the present wasteful use of "butcher's-meats," with positive advantage to the consumer. Many facts in support of this position will appear as we proceed. But another important object in furnishing the foregoing details is to point out how combinations of the nitrogenous, starchy, fatty, and mineral elements may be made, in well-proportioned mixtures, so as to produce what I have termed a "perfect" dish—perfect, that is, so far as the chief indication is concerned, viz., one which supplies every demand of the body, without containing any one element in undue proportion. For it is obvious that one or two of these elements may exist in injurious excess, especially for delicate stomachs, the varied peculiarities of which, as before insisted on, must sometimes render necessary a modification of all rules. Thus it is easy to make the fatty constituent too large, and thereby derange digestion—a result frequently experienced by persons of sedentary habits, to whom a little pastry, a morsel of *foie gras*, or a rich cream is a source of great discomfort, or of a "bilious attack;" while the laborer, who requires much fatty fuel for his work, would have no difficulty in consuming a large quantity of such compounds with advantage. Nitrogenous matter also is commonly supplied beyond the eater's wants; and if more is consumed than can be used for the purposes which such aliment subserves, it must be eliminated in some way from the system. This process of elimination, it suffices to say here, is undoubtedly a prolific cause of disease, such as gout and its allies, as well as other affections of a serious character, which would in all probability exist to a very small extent, were it not the habit of those who, being able to obtain the strong or butcher's-meats, eat them daily, year after year, in larger quantity than the constitution can assimilate.

Quitting the subject of wheat and the leguminous seeds, it will be interesting to review briefly the combinations of rice, which furnishes so large a portion of the world with a vegetable staple of diet. Remembering that it contains chiefly starch, with nitrogen in small proportion, and almost no fat or mineral elements, and just sufficing, perhaps, to meet the wants of an inactive population in a tropical climate, the first addition necessary for people beyond this limit will be fat, and, after that, more nitrogen. Hence the first effort to make a dish of rice "complete" is the addition of butter and a little Parmesan cheese, in the simple *risotto*, from which, as a starting-point, improvement, both for nutritive purposes and for the demands of the palate, may be carried to any extent. Fresh additions are made in the shape of marrow, of morsels of liver, etc., of meat broth with onion and spice, which constitute the mixture, when well prepared, nutritious and highly agreeable. The analogue of this mainly Italian dish is the *pilau*, or *pilaff*, of the Orientals, consisting, as it does, of nearly the same materials, but differently arranged. The curry of poultry and the kedgeree of fish are further varieties which it is unnecessary to describe. Follow the same combination to Spain, where we find a popular national dish, but slightly differing from the foregoing, in the *pollo con arroz*, which consists of abundance of rice, steeped in a little broth, and containing morsels of fowl, bacon, and sausage, with appetizing spices, and sufficing for an excellent meal.

Another farinaceous product of worldwide use is the maize or Indian corn. With a fair amount of nitrogen, starch, and mineral elements, it contains also a good proportion of fat, and is made into bread, cakes, and puddings of various kinds. It is complete, but susceptible of improvement by the addition of nitrogen. Hence, in the United States, where it is largely used, it is often eaten with beans under the name of "succotash." In Italy it is ground into the beautiful yellow flour which is conspicuous in the streets of almost every town; when made into a firm paste by boiling in water, and sprinkled with Parmesan cheese, a nitrogenous aliment, it becomes what is known as *polenta*, and is largely consumed with some relish in the shape of fried fish, sardines, sausage, little birds, or morsels of fowl or goose, by which of course fresh nitrogen is added. Macaroni has been already alluded to; although rich in nitrogenous and starchy materials, it is deficient in fat. Hence it is boiled and eaten with butter and parmesan (*à l'italienne*) and with tomatoes, which furnish saline matters, with meat gravy, or with milk.

Nearer home the potato forms a vegetable basis in composition closely resembling rice, and requiring therefore additions of nitrogenous and fatty elements. The Irishman's inseparable ally, the pig, is the natural, and to him necessary, complement of the tuber, making the latter a complete and palatable dish. The every-day combination of mashed potato and sausage is an application of the same principle. In the absence of pork, the potato eater substitutes a cheap oily fish, the herring. The combination of fatty material with the potato is still further illustrated in our baked potato and butter, in fried potatoes in their endless variety of form, in potato mashed with milk or cream, served in the ordinary way with *maître d'hôtel* butter, or arriving at the most perfect and finished form in the *pommes de terre sautées au beurre* of a first-class French restaurant, where it becomes almost a *plat de luxe*. Even the simple bread-and-butter or bread-and-cheese of our own country equally owe their form and popularity to physiological necessity; the deficient fat of the bread being supplemented by the fatty elements of each addition, the cheese supplying also its proportion of nitrogenous matter, which exists so largely in its peculiar principle, caseine. So, again, all the suet puddings, "shortcake," pie-crust, or pastry, whether baked or boiled, consist simply of farinaceous food rendered stronger nutriment by the addition of fatty matter.

In the same way almost every national dish might be analyzed up to the *pot-au-feu* of our neighbors, the right management of which combines nutritious quality with the abundant aroma and flavor of fresh vegetables which enter so largely into this economical and excellent mess.

It will be apparent that, up to this point,

our estimate of the value of these combinations has been limited, or almost so, by their physiological completeness as foods, and by their economical value in relation to the resources of that great majority of all populations, which is poor. But when the inexorable necessity for duly considering economy has been complied with, the next aim is to render food as easily digestible as possible, and agreeable to the senses of taste, smell, and sight.

The hard laborer with simple diet, provided his aliment is complete and fairly well cooked, will suffer little from indigestion. He cannot be guilty, for want of means, of eating too much, fertile source of deranged stomach with those who have the means; physical labor being also in many circumstances the best preventive of dyspepsia. "Live on sixpence a day and earn it," attributed to Abernethy as the sum of his dietary for a gluttonous eater, is a maxim of value, proved by millions. But for the numerous sedentary workers in shops, offices, in business and professions of all kinds, the dish must not only be "complete;" it must be so prepared as to be easily digestible by most stomachs of moderate power, and it should also be as appetizing and agreeable as circumstances admit.

On questioning the average middle-class Englishman as to the nature of his food, the all but universal answer is, "My living is plain, always roast and boiled"—words which but too clearly indicate the dreary monotony, not to say unwholesomeness, of his daily food; while they furthermore express his satisfaction, such as it is, that he is no luxurious feeder, and that, in his opinion, he has no right to an indigestion. Joints of beef and mutton, of which we all know the very shape and changeless odors, follow each other with unvarying precision, six roast to one boiled, and have done so ever since he began to keep house some five-and-twenty years ago! I am not sanguine enough to suppose that this unbroken order which rules the dietary of the great majority of British families of moderate and even of ample means, will be disturbed by any suggestions of mine. Nevertheless, in some younger households, where habits, followed for want of thought or knowledge, have not yet hardened into law, there may be a disposition to adopt a healthier diet and a more grateful variety of aliment. For variety is not to be obtained in the search for new animal food. Often as the lament is heard that some new meat is not discovered, that the butcher's display of joints offers so small a range for choice, it is not from that source that wholesome and pleasing additions to the table will be obtained.

But our most respectable paterfamilias, addicted to "plain living," might be surprised to learn that the vaunted "roasting" has no longer in his household the same significance it had five-and-twenty years ago; and that probably, during the latter half of that term, he has eaten no roast meat, whatever he may aver to the contrary. Baking, at best in a half-ventilated oven, has long usurped the function of the spit in most houses, thanks to the ingenuity of economical range-makers. And the joint, which formerly turned in a current of fresh air before a well-made fire, is now half stifled in a close atmosphere of its own vapors, very much to the destruction of the characteristic flavor of a roast. This is a smaller defect, however, as regards our present object, than that which is involved by the neglect in this country of braising as a mode of cooking animal food. By this process more than mere "stewing" is of course intended. In braising, the meat is just covered with a strong liquor of vegetable and animal juices (*braise* or *mirepoix*) in a closely covered vessel, from which as little evaporation as possible is permitted, and is exposed for a considerable time to a surrounding heat just short of boiling. By this treatment tough fibrous flesh, whether of poultry or of cattle, or meat unduly fresh, such as can alone be procured during the summer season in towns, is made tender, and is furthermore impregnated with the odors and flavor of fresh vegetables and sweet herbs. Thus, also, meats which are dry, or of little flavor, as veal, become saturated with juices and combined with sapid substances, which render the food succulent and delicious to the palate. Small portions sufficing for a single meal, however small the family, can be so dealt with; and a *réchauffée,* or cold meat for to-morrow, is not a thing of necessity, but only of choice when preferred. Neither time nor space permits me to dwell further on this topic, the object of this paper being rather to furnish suggestions than explicit instruction in detail.

The art of frying is little understood, and the omelet is almost entirely neglected by our countrymen. The products of our frying-pan are often greasy, and therefore, for many persons, indigestible, the shallow form of the pan being unsuited for the process of boiling in oil, that is, at a heat of nearly 500° Fahr., that of boiling water being 212°. This high temperature produces results, which are equivalent indeed to quick roasting, when the article to be cooked is immersed in the boiling fat. Frying, as generally conducted, is rather a combination of broiling, toasting, or scorching; and the use of the deep pan of boiling oil or dripping, which is essential to the right performance of the process, and especially preventing greasiness, is a rare exception, and not the rule, in ordinary kitchens. Moreover, few English cooks can make a tolerable omelet; and thus one of the most delicious and nutritious products of culinary art, with the further merit that it can be more rapidly prepared than any other dish, must really at present be regarded as an exotic. Competent instruction at first and a little practice are required, in order to attain a mastery in producing an omelet; but these given, there is no difficulty in turning out a first-rate specimen. The ability to do this may be so useful in the varied circumstances of travel, etc., that no young man destined for foreign service, or even who lives in chambers, should fail to attain the easily acquired art.

The remainder of the second portion of my subject—viz., the preparation of food—must appear, although in very brief terms, at the commencement of this paper. After which I shall proceed to consider the chief object of the present article, viz., the combination and service of dishes to form a meal—especially in relation to dinners and their adjuncts.

I think it may be said that soups, whether clear (that is, prepared from the juices of meat and vegetables only), or thick (that is, *purées* of animal or vegetable matters), are far too lightly esteemed by most classes in England, while they are almost unknown to the working-man. For the latter they might furnish an important cheap and savory dish; by the former they are too often regarded as the mere prelude to a meal, to be swallowed hastily, or disregarded altogether as mostly unworthy of attention. The great variety of vegetable *purées*, which can be easily made and blended with light animal broths, admits of daily change in the matter of soup to a remarkable extent, and affords scope for taste in the selection and combination of flavors. The use of fresh vegetables in abundance —such as carrots, turnips, artichokes, celery, cabbage, sorrel, leeks, and onions—renders such soups wholesome and appetizing.

The supply of garden produce ought, in this country, to be singularly plentiful; and, owing to the unrivalled means of transport, all common vegetables ought to be obtained fresh in every part of London. The contrary, however, is unhappily the fact. It is a matter of extreme regret that vegetables, dried and compressed after a modern method, should be so much used as they are for soups, by hotel-keepers and other caterers for the public. Unquestionably useful as these dried products are on board ship and to travellers camping out, to employ them at home, when fresh can be had, is the result of sheer indolence or of gross ignorance. All the finest qualities of scent and flavor, with some of the fresh juices, are lost in the drying process; and the infusions of preserved vegetables no more resemble a freshly made odoriferous soup, than a cup of that thick brown, odorless, insipid mixture, consisting of some bottled "essence" dissolved in hot water, and now supplied as coffee at most railway stations and hotels in this country, resembles the recently made infusion of the freshly-roasted berry. It says little for the taste of our countrymen that such imperfect imitations are so generally tolerated without complaint.

The value of the gridiron is, perhaps, nowhere better understood than in England, especially in relation to chops, steak, and kidney. Still, it is not quite so widely appreciated as it deserves to be in the preparation of many a small dish of fish, fowl, and meat, to say nothing of a grilled mushroom, either alone or as an accompaniment to any of them. And it may be worth while, perhaps, remarking that the sauce *par excellence* for broils is mushroom catsup; and the garnish cool lettuce, watercress, or endive. And this suggests a word or two on the important addition which may be made to most small dishes of animal food under the title of "garnish." Whether it be a small fillet, braised or roasted, or a portion thereof broiled; a fricandeau, or the choice end of a neck of mutton made compact by shortening the bones; or a small loin, or a dish of trimmed neck cutlets, or a choice portion of broiled rumpsteak; a couple of sweetbreads, poultry, pigeon, or what not—the garnish should be a matter of consideration. Whether the dish be carved on the family table, as it rarely fails to be when its head is interested in the cuisine, or whether it is handed in the presence of guests, the quality and the appearance of the dish greatly depend on the garnish. According to the meat may be added, with a view both to taste and appearance, some of the following: *purées* of sorrel, spinach, and other greens, of turnips, and of potatoes plain, in shapes, or in croquettes; cut carrots, pease, beans, endive, sprouts, and other green vegetables; stewed onions, small or Spanish; cucumbers, tomatoes, macaroni in all forms; sometimes a few sultanas boiled, mushrooms, olives, truffles. In the same way chestnuts are admirable, whole, boiled or roasted, and as a *purée* freely served, especially in winter when vegetables are scarce; serving also as farce for fowls and turkeys. While such vegetables as green pease, French and young broad beans, celery and celeriac, asparagus, sea-kale, cauliflower, spinach, artichokes, vegetable marrows, etc., are worth procuring in their best and freshest condition, to prepare with special care as separate dishes.

It is doubtful whether fish is esteemed so highly as an aliment as its nutritious qualities entitle it to be, while it offers great opportunity for agreeable variety in treatment. As a general observation, it may be said that in preparing it for table sufficient trouble is not taken to remove some portion of the bones; this can be advantageously done by a clever cook without disfiguring or injuring the fish. Sauces should be appropriately served : for example, the fat sauces, as *hollandaise* and other forms of melted butter, are an appropriate complement of hot boiled fish, while *mayonnaise* is similarly related to cold. These and their variations, which are numerous, may also accompany both broiled and fried fish, but these are often more wholesome and agreeable when served with only a squeeze of lemon-juice, and a few grains of the zest, if approved, when a fresh green lemon is not to be had—and it rarely can be here.

But the juice of the mushroom is preferred, and no doubt justly, by some. Endless variations and additions may be made according to taste on these principles. But there is another no less important principle, viz., that the fish itself often furnishes a sauce from its own juices, more appropriate than some of the complicated and not very digestible mixtures prepared by the cook. Thus "melted butter"—which is regarded as essentially an English sauce—when intended to accompany fish, should not be, as it almost invariably is, a carelessly made compound of butter, flour, and water; but in place of the last-named ingredient there should be a concentrated liquor made from the trimmings of the fish itself, with the addition of a few drops of lemon-juice, and strengthened, if necessary, from other sources, as from shell-fish of some kind. Thus an every-day sauce of wholesome and agreeable quality is easily made: it finds its highest expression in that admirable dish, the sole, with *sauce au vin blanc* of the French, or, as associated with shell-fish, in the *sole à la normande*. Some

FOOD AND FEEDING. 71

fish furnish their own sauce in a still simpler manner, of which an illustration no less striking is at hand in the easiest but best mode of cooking a red mullet, viz., baking it, and securing the gravy of delicious flavor, which issues abundantly from the fish, chiefly from the liver, as its only sauce.

Passing rapidly on without naming the ordinary and well-known service of cold meats, fresh and preserved, poultry and game, open or under paste in some form, to be found in profusion on table or sideboard, and in which this country is unrivalled, a hint or two relating to some lighter cold *entrées* may be suggested. It is scarcely possible to treat these apart from the salad which, admirable by itself, also forms the natural garnish for cold dishes. A simple aspic jelly, little more than the *consommé* of yesterday flavored with a little lemon-peel and tarragon vinegar, furnishes another form of garnish, or a basis for presenting choice morsels in tempting forms, such as poultry livers, ox-palates, quenelles, fillets of game, chicken, wild-fowl, fish, prawns, etc., associated with a well-made salad. On this system an enterprising cook can furnish many changes of light but excellent nutritious dishes.

On salad so much has been written, that one might suppose, as of many other culinary productions, that to make a good one was the result of some difficult and complicated process, instead of being simple and easy to a degree. The materials must be secured fresh, are not to be too numerous and diverse, must be well cleansed and washed without handling, and all water removed as far as possible. It should be made by the hostess, or by some member of the family, immediately before the meal, and be kept cool until wanted. Very few servants can be trusted to execute the simple details involved in cross-cutting the lettuce, endive, or what not, but two or three times in a roomy salad-bowl; in placing one salt-spoonful of salt and half that quantity of pepper in a table-spoon, which is to be filled three times consecutively with the best fresh olive oil, stirring each briskly until the condiments have been thoroughly mixed, and at the same time distributed over the salad. This is next to be tossed well, but lightly, until every portion glistens, scattering meantime a little finely chopped fresh tarragon and chervil, with a few atoms of chives over the whole. Lastly, but only immediately before serving, one small table-spoonful of mild French vinegar is to be sprinkled over all, followed by another tossing of the salad.[1] The uncooked tomato, itself the prince of salads, may be sliced and similarly treated for separate service, or added to the former, equally for taste and appearance. Cold boiled asparagus served with a *mayonnaise* forms a dish of its kind not to be surpassed. At present ranking, when the quality is fine, as an expensive luxury, there is no reason why, with the improved methods of cultivating this delicious and wholesome vegetable, it should not be produced in great abundance, and for less than half its present price.[2] As to the manifold green stuffs

---

[1] A salad for five or six persons is supposed.

[2] On this subject, and also on salad culture, see *The Parks and Gardens of Paris*, by W. Robinson, F.L.S., p. 468 et seq. 2d ed. Macmillan, 1878.

which, changing with the season, may be presented as salad, their name is legion; and their choice must be left to the eater's judgment, fancy, and digestion, all of which vary greatly.

The combination of dishes to form a meal now demands our consideration. The occupations of man in a civilized state, no less than the natural suggestions of his appetite, require stated and regular times for feeding. But the number of these set apart in the twenty-four hours differs considerably among different peoples and classes. Taking a general view of the subject, it may be said that there are three principal systems to which all varieties of habit may be reduced. From an English point of view, these may be regarded as—

1. The Continental system of two meals a day.
2. The system of provincial life (Great Britain), or four meals.
3. The system of town life (ditto), or three meals.

(1.) In the Continental system, the slight

refreshment served in the early morning, in the form of coffee or chocolate, with a rusk or a morsel of bread, does not amount to a meal. It is only a dish, and that a light one, and not a combination of dishes, which is then taken. At or about noon a substantial meal, the déjeûner, is served; and at six or seven o'clock an ample dinner. Such is the two-meal system, and it appears to answer well throughout the west and south of Europe.

(2.) What I have termed the provincial system consists of a substantial breakfast at eight or nine, a dinner at one or two, a light tea about five, and a supper at nine or ten. It is this which is popular throughout our own provincial districts, and also among middle-class society of our northern districts throughout both town and country. The habits also of the great German nation correspond more to this than to the first-named system.

(3.) The prevailing system of London, and of the numerous English families throughout the country whose habits are formed from partial residence in town, or by more or less intimate acquaintance with town life, is that of three meals daily. In general terms the breakfast takes place between eight and ten; the lunch from one to two; the dinner from half-past six to eight.

In all cases each meal has its own specific character. Thus, here, breakfast is the most irregular in its service, and least of all demands general and intimate coherence of the party assembled. Individual interests concerned in the letter-bag, in the morning news, in plans for the day, in cares of coming business, etc., are respected. Provision for acknowledged dietetic peculiarities on the part of individuals is not forgotten, and every one comes or goes as he pleases.

At lunch the assembly is still somewhat uncertain. Thus some members of the family are absent without remark; intimate friends may appear without special invitation; while those less intimate can be asked with small ceremony. Occupations of pleasure or of business still press for pursuit during the afternoon, and the meal for such may not be too substantial. It should suffice amply to support activity; it should never be so considerable as to impair it.

The last meal of the three—dinner—has characters wholly different from the preceding. The prime occupations of the day are over; the guests are known and numbered; the sentiment is one of reunion after the dispersion of the day—of relaxation after its labors, sports, or other active pleasures. Whatever economy of time may have been necessary in relation to the foregoing meals, all trace of hurry should disappear at dinner. A like feeling makes the supper of the "provincial" system a similarly easy and enjoyable meal. And all this is equally true of dinner, whether it unites the family only, or brings an addition of guests. General conversation—the events and personal incidents of the day, the current topics of the hour, are discussed in a light spirit, such as is compatible with proper attention to the dishes provided. All that follows late dinner should for the most part be amusement—it may be at the theatre, an evening party, or a quiet evening at home. There should be ample time, however, for every coming engagement, and security for some intervening rest for digestion. Dinner, then, is the only meal which — as the greater includes the less — need be discussed in the third part of our subject, which claims to treat of custom and art in combining dishes to form a repast. With the requirements and under the circumstances just specified, it should not be a heavy meal, but it should be sufficing. No one after dinner should feel satiety or repletion, with a sense of repugnance at the idea of eating more; but all should still enjoy the conviction that a good meal furnishes delightful and refreshing occupation.

Dinners are of two kinds—the ordinary meal of the family, and the dinner to which guests are invited. There is a third dinner in this country, of common—too common—occurrence, viz., the public dinner, which is essentially a British institution, and cannot be passed by in silence.

The late dinner should never include children. It is a meal which is in every way unsuited to them, and they are quite unfitted to take part in its functions; besides, the four-meal system is better adapted to their requirements of growth and digestion in early life. A family dinner may usually consist of a soup, fish, *entrée*, roast and sweet; the *entrée* may even be omitted; on the other hand, if the meal is required to be more substantial, a joint may be served in addition after the fish; but this should be very rarely necessary. A dish of vegetables may be advantageously placed before or after the roast, according to circumstances; and supplementary vegetables should be always at hand.

The *rationale* of the initial soup has often been discussed: some regard it as calculated to diminish digestive power, on the theory that so much fluid taken at first dilutes the gastric juices. But there appears to be no foundation for this belief: a clear soup, or the fluid constituents of a *purée*, disappear almost immediately after entering the stomach, being absorbed by the proper vessels, and in no way interfere with the gastric juice which is stored in its appropriate cells ready for action. The habit of commencing dinner with soup has without doubt its origin in the fact that aliment in this fluid form—in fact, ready digested— soon enters the blood and rapidly refreshes the hungry man, who, after a considerable fast and much activity, sits down with a sense of exhaustion to commence his principal meal. In two or three minutes after taking a plate of good warm *consommé*, the feeling of exhaustion disappears, and irritability gives way to the gradually rising sense of good-fellowship with the circle. Some persons have the custom of allaying exhaustion with a glass of sherry before food—a gastronomic no less than a physiological blunder, injuring the stomach and depraving the palate. The soup introduces at once into the system a small instalment of ready-digested food, and saves the short period of time which must be spent by the stomach in deriving some portion of nutriment from solid aliment, as well as indirectly strengthening the organ of digestion itself for its forthcoming duties. Few will be found to dispute the second place in order to fish, although this arrangement is in some quarters an open question: its discussion, however, can scarcely be regarded as within the limit of our space. The third dish should consist of the chief meat, the joint, if desired; if not, one of the smaller dishes of meat, such as fricandeau, cutlets, fillet, or sweetbread, before spoken of, well garnished, will be appropriate, and to many preferable. Next the well-roasted bird—of game or poultry —accompanied or followed by salad, and a dish of choice vegetables. Then one light, simple sweet, for those who take it, and a slight savory biscuit or morsel of cheese, completes the repast. Such a meal contains within its limits all that can be desired for daily enjoyment and use. If well and liberally served, it is complete in every sense of the word. Dessert and its extent is a matter of individual taste; of wines, coffee, and liqueurs I shall speak hereafter.

A word about *hors d'œuvres*. It is well known that the custom exists, to a very wide extent, among Continental nations of commencing either mid-day déjeûner or dinner by eating small portions of cold pickled fish, vegetables, of highly-flavored sausage thinly sliced, etc., to serve, it is said, as a whet to appetite. This custom reaches its highest development in the *zakuska* of the Russian, which, consisting of numerous delicacies of the kind mentioned, is sometimes to be found occupying a table in an anteroom to be passed between the drawing-room and dining-room; or, and more commonly, spread on the sideboard of the latter. The Russian eats a little from three or four dishes at least, and "qualifies" with a glass of strong grain spirit (*vodka*) or of some liqueur before taking his place at the table. Among these savory preliminaries may often be found caviare in its fresh state, gray, pearly, succulent, and delicate, of which most of the caviare found in this country is, speaking from personal experience of both, but as the shadow to the substance.

I have no hesitation in saying, after much consideration of the practice of thus commencing a meal, that it has no *raison d'être* for persons with healthy appetite and digestion. For them, both pickled food and spirit are undesirable, at any rate on an empty stomach. And the *hors d'œuvres*, although attempts to transplant them here are often made, happily do not, as far as I have observed, thrive on our soil. They have been introduced here chiefly, I think, because their presence, being demanded by foreign gastronomic taste, is supposed to be therefore necessarily correct. But the active exercise and athletic habits of the Englishman, his activity of body and mind in commercial pursuits, all tend to bring him to the dinner-table wanting food rather than appetite, and in no mind to ask for "whets" to increase it. Among idle men, whose heavy lunch, liberally accompanied with wine, and not followed by exercise, has barely disappeared from the stomach at the hour of dinner, a piquant prelude as stimulus of appetite is more appreciated. Hence the original invention of *hors d'œuvres;* and their appearance in a very much slighter and more delicate form than that which has been described, still to be observed in connection with the chief repasts of the Latin races. The one plate which heralds dinner, indigenous to our country, is also one of its own best products — the oyster. But this is scarcely a *hors d'œuvre.* In itself a single service of exquisite quality, served with attendant graces of delicate French vinegar, brown bread and butter, and a glass of light chablis for those who take it, the half-dozen natives occupying the hollow shells, and bathed in their own liquor, hold rank of a very different kind to that of the miscellaneous assortment of tidbits alluded to. Oysters are, in fact, the first dish of dinner, and not its precursor; the first chapter, and not the advertisement. And this brings us to the dinner of invitation.

And of this dinner there are two very distinct kinds. First there is the little dinner of six or eight guests, carefully selected for their own specific qualities, and combined with judgment to obtain an harmonious and successful result. The ingredients of a small party, like the ingredients of a dish, must be well chosen to make it "complete." Such are the first conditions to be attained in order to achieve the highest perfection in dining. Secondly, there is the dinner of society, which is necessarily large; the number of guests varying from twelve to twenty-four.

The characteristics of the first dinner are —comfort, excellence, simplicity, and good taste. Those of the second are—the conventional standard of quality, some profusion of supply, suitable display in ornament and service.

It must be admitted that, with the large circle of acquaintances so commonly regarded as essential to existence in modern life, large dinners only enable us to repay our dining debts, and exercise the hospitality which position demands. With a strong preference, then, for the little dinners, it must be admitted that the larger banquet is a necessary institution; and therefore we have only to consider now how to make the best of it.

No doubt the large dinner has greatly improved of late; but it has by no means universally arrived at perfection. Only a few years ago excellence in quality and good taste in cuisine were often sacrificed in the endeavor to make a profuse display. Hence, abundance without reason, and combinations without judgment, were found co-existing with complete indifference to comfort in the matters of draughts, ventilation, temperature, and consumption of time. Who among the diners-out of middle age has not encountered many a time an entertainment with some such programme as the following: one of an order which, it is to be feared, is not even yet quite extinct?

Eighteen or twenty guests enter a room adapted at most to a dinner of twelve. It is lighted with gas; the chief available space being occupied by the table, surrounding which is a narrow lane, barely sufficing for the circulation of the servants. Directly — perhaps after oysters — appear turtle-soups, thick and clear. A *consommé* is to be had on demand, but so unexpected a choice astonishes the servitor, who brings it after some delay, and cold: with it, punch. Following, arrive the fish — salmon and turbot, one or both, smothered in thick lobster sauce: sherry. Four *entrées* promenade the circuit in single file, where-of the first was always oyster patties; after which came mutton or lamb cutlets, a vol-au-vent, etc.: hock and champagne. Three-quarters of an hour at least, perhaps an hour, having now elapsed, the saddle or haunch of mutton arrives, of which gentlemen who have patiently waited get satisfactory slices, and currant-jelly, with cold vegetables or a heavy flabby salad. Then come boiled fowls and tongue, or a turkey with heavy force-meat; a slice of ham, and so on, up to game, followed by hot substantial pudding, three or four other sweets, including an iced pudding; wines in variety, more or less appropriate; to be followed by a *pâté de foie gras*, more salad, biscuits and cheese. Again, two ices, and liqueurs. Then an array of decanters, and the first appearance of red wine; a prodigious dessert of all things in and out of season, but particularly those which are out of season, as being the more costly. General circulation of waiters, handing each dish in turn to everybody, under a running fire of negatives, a ceremonial of ten or fifteen minutes' duration, to say the least.

Circulation of decanters; general rustle of silks, disappearance of the ladies; and first change of seat, precisely two hours and a half after originally taking it. It may be hoped that a charming companion on either side has beguiled and shortened a term which otherwise must have been felt a little long. Now the general closing up of men to host, and reassembling of decanters; age and qualities of wine, recommendation of vintages. Coffee which is neither black nor hot. Joining the ladies; service of gunpowder-tea, fatal to the coming night's rest if taken in a moment of forgetfulness; and carriages announced.

Admitted that such an exhibition is impossible now in any reasonable English circle, it nevertheless corresponds very closely in style with that of the public dinner; a state of things without excuse. And the large private dinner is still generally too long, the menu too pretentious. Let me, however, be permitted to record, equally in proof of growing taste and as grateful personal duty, how many admirable exceptions to the prevailing custom are now afforded. Then, of course, it must be understood that while the dinner for six or eight persons is designed as an harmonious whole of few, well-chosen dishes, all of which are intended to be eaten in their order, the menu of the larger party must offer various dishes for choice to meet the differing tastes of more numerous guests, and it must therefore be larger. Let us see how this is to be met. First, the soups: it is the custom to offer a *consommé*, which ought to be perfect in clearness, color, and savor, and to be served perfectly hot; containing vegetables, etc., variously treated — doubtless the best commencement, as it is the key-note, of the dinner; revealing also, as it does nine times out of ten, the calibre of the cook to whose talent the guest is intrusted. But there is mostly an alternative of "white soup," and this is almost always a mistake. Many persons refuse it, and they are right, containing, as it generally does, a considerable proportion of cream — an injudicious beginning, when there is much variety to follow; excellent sometimes as one of three or four dishes, but dangerous otherwise to the guest who has not an exceptionally powerful digestion. But suppose oysters, vinegar, and chablis have just been swallowed! A brown *purée*, as of game, or one of green vegetable, less frequently met with, would be far safer. Two fish, of course, should always be served; as, for example, a slice of Severn or Christchurch salmon, just arrived from the water, for its own sake; and a fillet of white-fish for the sake of its sauce and garnish, which should be therefore perfect. The next dish is, in London, a question under discussion: viz., the question of precedence to an *entrée*, or to the *pièce de résistance*. The custom has been to postpone the appearance of the latter until lighter dishes have been despatched or declined. If, however, the English joint is required at a meal already comprehensive in the matter of dishes, and taken at a late hour, it seems more reasonable to serve it next to the fish, when those who demand a slice of meat may be expected to have an appropriate appetite, which will certainly be impaired, equally by accepting the *entrées*, or fasting partially without them. After the joint, two light *entrées* may follow, and these must necessarily be either in themselves peculiarly tempting morsels, or products of culinary skill, offering inducement to the palate rather than to an appetite which is no longer keen. Then the best roast possible in season, and a salad; a first-rate vegetable, two choice sweets, one of which may be iced; a light savory biscuit or a morsel of fine barely salted caviare, which may be procured in one or two places at most in town, will complete the dinner. For dessert, the finest fruits in season to grace the table and for light amusement after; or simply nuts in variety, and dry biscuits; nothing between the two is tolerable, and little more than the latter is really wanted; only for decorative purposes fruit equals flowers. But it may be admitted that the diminished number of sweet *entremets* strengthens the plea for a supply of delicious fruits, rendering the dessert useful and agreeable as well as ornamental.

And now that dessert is over, let me say that I do not admit the charge sometimes intimated, although delicately, by foreigners, of a too obvious proclivity to self-indulgence on the part of Englishmen, in permitting the ladies to leave the table without escort to the drawing-room. The old custom of staying half an hour, or even an hour afterward, to drink wine, which is doubtless a remnant of barbarism, has long been considered indefensible. Still, the separation of the party into two portions for fifteen or twenty minutes is useful to both, and leads, perhaps, more completely to a general mixture of elements on re-union after than is attained by the return of the original pairs together. Whether this be so or not, the ladies have a short interval for the interchange of hearsays and ideas relative to matters chiefly concerning their special interests; while the men enjoy that indispensable finish to a good dinner, an irreproachable cup of coffee and a cigarette, and the sooner they arrive the better. With the small dinners of men it can scarcely too quickly follow the last service.

But marked by a special character are

some dinners, which may be either small or large in relation to the number of guests, but which are necessarily limited as regards the variety of aliments served. I refer to dinners at which either turtle or fish predominate. In accordance with a principle already enunciated, a bowl of substantial stock, containing four or five broad flakes of the gelatinous product, often miscalled "fat," which alone represents the turtle in the compound, is not a judicious prelude to a dinner arranged according to the orthodox programme, and offering the usual variety. A lover of turtle indulges freely in the soup, both thick and clear, making it in fact an important instalment of his repast; and he desires, with or without some slight interlude, to meet the favorite food again in the form of an *entrée*. After so substantial a commencement, the dinner should be completed chiefly by poultry, and game if in season, and for the most part by dishes which are grilled or roast, in contrast to the succulent morsels which have preceded.

The fish dinner, also an occasional departure from daily routine, is acceptable, and gratifies the taste for that delicate and pleasant food in considerable variety. But if so indulged, very few dishes ought to appear subsequently. It is a curious fact that the traditional bacon and beans, which appear toward the close of a Greenwich whitebait dinner, should afford another illustration of undesigned compliance with the natural law referred to at the outset, the bacon furnishing complementary fat to supply its notable absence in fish.

The enjoyment of a curry—and when skilfully made it is almost universally admitted to be one of the most attractive combinations which can be offered to the senses of taste and smell — is only possible at a limited repast. When freely eaten, very little is acceptable to the palate afterward, exhausted as it is by the pervading fragrance of the spice and other adjuncts. Hence a curry should form the climax of a short series of dishes leading up to it: when presented, as it sometimes is, among the *entrées* of a first course, it is wholly out of place.

Here we may appropriately take a rapid glance at the characteristics of the feast where the guests are few in number.

The small dinner-party should be seated at a round or oval table, large enough for personal comfort, small enough to admit of conversation in any direction without effort. The table should of course be furnished with taste, but is not to be encumbered with ornaments, floral or other, capable of obstructing sight and sound. A perfect *consommé*, a choice of two fish, a *filet* or a châteaubriand, a *gigot* or a fricandeau; followed by a chaudfroid, a *crème de volaille garni*, a roast and salad, a choice vegetable, and an iced *soufflé* or *charlotte;* and in summer a *macédoine* of fresh fruits in an old china family bowl, if there is one; and lastly, a savory biscuit; accompanying vegetables and appropriate wines; may be regarded as furnishing a scheme for such a party, or a theme of which the variations are endless. Seven or eight guests can thus be brought into close contact: with a larger number the party is apt to form two coteries, one on each side of the host. The number is a good one also in relation to the commissariat department—eight persons being well supplied by an *entrée* in one dish; while two are necessary for ten or twelve. Moreover, one bottle of wine divides well in eight; if, therefore, the host desire to give with the roast one glass of particularly fine ripe Corton or Pomard, a single bottle is equal to the supply; and so with any other choice specimen of which a single circulation is required; and of course the rule holds equally if the circuit is to be repeated.

And this leads us to the question—and an important one it is—of the Wine.

I have already said that, among all civilized nations, wine in some form has for centuries been highly appreciated as a gastronomic accompaniment to food. I cannot, and do not, attempt to deny it this position. Whether such employment of it is advantageous from a dietetic or physiological point of view is altogether another question. I am of opinion that the *habitual* use of wine, beer, or spirits is a dietetic error, say, for nineteen persons out of twenty. In other words, the great majority of the people, at any age or of either sex, will enjoy better health, both of body and mind, and will live longer, without any alcoholic drinks whatever, than with habitual indulgence in their use, even although such use be what is popularly understood as moderate. But I do not aver that any particular harm results from the habit of now and then enjoying a glass of really fine pure wine—and, rare as this is, I do not think any other is worth consuming—just as one may occasionally enjoy a particularly choice dish; neither the one nor the other, perhaps, being sufficiently innocuous or digestible for frequent, much less for habitual use. Then I frankly admit that there are some persons—in the aggregate not a few —who may take small quantities of genuine light wine or beer with very little if any appreciable injury. For these persons such drinks may be put in the category of luxuries permissible within certain limits or conditions; and of such luxuries let tobacco-smoking be another example. No one, probably, is any better for tobacco; and some people are undoubtedly injured by it; while others find it absolutely poisonous, and cannot inhale even a small quantity of the smoke without instantly feeling sick or ill. And some few indulge the moderate use of tobacco all their lives without any evil effects, at all events that are perceptible to themselves or to others.

Relative to these matters, every man ought to deal carefully and faithfully with himself, watching rigorously the effects of the smallest license on his mental and bodily states, and boldly denying himself the use of a luxurious habit if he finds any signs of harm arising therefrom. And he must perform the difficult task with a profound conviction that his judgment is very prone to bias on the side of indulgence, since the luxurious habit is so agreeable, and to refrain therefrom, in relation to himself and to the present opinion of society, so difficult. Be it remarked, however, that the opinion of society is notably and rapidly changing relative to the point in question.

Having premised thus much, I have only

now to say, first, that wine, in relation to dinner, should be served during the repast; it should never be taken, in any form or under any circumstances, before, that is, on an empty stomach, and rarely after the meal is finished. Regarded from a gastronomic point of view alone, nothing should appear after fruit but a small glass of cognac or liqueur, and coffee. The postprandial habit of drinking glass after glass even of the finest growths of the Gironde, or of the most mature or mellow shipments from Oporto, is doubtless a pleasant, but, in the end, for many persons, a costly indulgence.

Secondly, whatever wine is given should be the most sound and unsophisticated of its kind which can be procured. The host had far better produce only a bottle or two of sound *bourgeois* wine from Bordeaux — and most excellent wine may be found under such a denomination — with no pretence of a meretricious title, or other worthless finery about it, than an array of fictitious mixtures with pretentious labels procured from an advertising cheap winehouse. I could only speak in terms of contempt and disgust, did I not feel pity for the deluded victims, of the unscrupulous use of the time-honored and historical titles which advertisers shamelessly flaunt on bottles of worthless compounds by means of showy labels, in lists and pamphlets of portentous length, and by placards sown broadcast through the country. So that one may buy "Lafitte" or "Margaux," "Chambertin" or "Nuits," '47 port, or even '34, at any village store! No terms can be too strong to characterize such trade.

If fine wines of unquestionable character and vintage are to be produced, there are only two ways of possessing them: one, by finding some wine-merchant of long standing and reputation who will do an applicant the favor to furnish them, and the price must be large for quality and age. We may be certain that such a one will never advertise: no man who really has the *grands vins* of esteemed vintages in his cellar need spend a shilling in advertisements, for he confers a favor on his customer by parting with such stock. But better and more satisfactory is it to obtain from time to time a piece or two of wine, of high character and reputed vintage, when they are to be had, just fit to bottle, and lay them down for years until ripe for use. Commencing thus in early life, a man's cellar becomes in twenty or thirty years a possession of interest and value, and he can always produce at his little dinners, for those who can appreciate it, something curiously fine, and free, at all events, from the deleterious qualities of new and fictitious wines.

Briefly: the rule, by general gastronomic consent, for those who indulge in the luxury of wine, is to offer a glass of light pale sherry or dry Sauterne after soup; a delicate Rhine wine, if required, after fish; a glass of Bordeaux with the joint of mutton; the same, or champagne — dry, but with some true vinous character in it, and not the tasteless spirit-and-water just now enjoying an evanescent popularity — during the *entrées;* the best red wine in the cellar, Bordeaux or Burgundy, with the grouse or other roast game; and— But this ought to suffice, even for that exceptional individual who is supposed to be little if at all injured by "moderate" potations. With the ice or dessert, a glass of full-flavored but matured champagne, or a liqueur, may be served; but at this point dietetic admonitions are out of place, and we have already sacrificed to luxury. The value of a cigarette at this moment is that with the first whiff of its fragrance the palate ceases to demand either food or wine. After smoke the power to appreciate good wine is lost, and no judicious host cares to open a fresh bottle from his best bin for the smoker, nor will the former be blamed by any man for a disinclination to do so.

For unquestionably tobacco is an ally of temperance; certainly it is so in the estimation of the gourmet. A relationship for him of the most perfect order is that which subsists between coffee and fragrant smoke. While wine and tobacco are antipathetic, the one affecting injuriously all that is grateful in the other, the aroma of coffee "marries" perfectly with the perfume of the finest leaf. Among the Mussulmans this relationship is recognized to the fullest extent; and also throughout the Continent the use of coffee, which is almost symbolical of temperate habits, is intimately associated with the cigarette or cigar. Only by the uncultured classes of Great Britain and of other Northern nations, who appear to possess the most insensitive palates in Europe, have smoke and alcoholic drinks been closely associated. By such, tobacco and spirit have been sought chiefly as drugs, and are taken mainly for their effects on the nervous system—the easy but disastrous means of becoming stupid, besotted, or drunk. People of cultivated tastes, on the other hand, select their tobacco or their wines, not for their qualities as drugs, but for those subtler attributes of flavor and perfume, which exist often in inverse proportion to the injurious narcotic ingredients; which latter are as much as possible avoided, or are accepted chiefly for the sake of the former.

Before quitting the subject of dining, it must be said that, after all, those who drink water with that meal probably enjoy food more than those who drink wine. They have generally better appetite and digestion, and they certainly preserve an appreciative palate longer than the wine-drinker. Water is so important an element to them, that they are not indifferent to its quality and source. As for the large class which cannot help itself in this matter, the importance of an ample supply of uncontaminated water cannot be overrated. The quality of that which is furnished to the population of London is inferior, and the only mode of storing it possible to the majority renders it dangerous to health. Disease and intemperance are largely produced by neglect in relation to these two matters. It would be invidious, perhaps, to say what particular question of home or foreign politics could be spared, that Parliament might discuss a matter of such pressing urgency as a pure water supply; or to specify what particular part of our enormous expenditure, compulsory and voluntary, might be better employed than at present, by diverting a portion to the attainment of that end. But for those who can afford to buy water, no purer exists in any natural sources than that of our own Malvern springs, and these are aërated and provided in the form of soda and potash waters of unexceptionable quality. Pure water, charged with gas, does not keep so long as a water to which a little soda or potash is added; but for this purpose six to eight grains in each bottle suffice — a larger quantity is undesirable. All the great makers of these beverages have now their own artesian wells or other equally trustworthy sources, so that English aërated waters are unrivalled in excellence. On the other hand, the foreign *siphon*, made, as it often is, at any chemist's shop, and from the water of the nearest source, is a very uncertain production. Probably our travelling fellow-countrymen owe their attacks of fever more to drinking water contaminated by sewage matter than to the malarious influences which pervade certain districts of Southern Europe. The only water safe for the traveller to drink is a natural mineral water, and such is now always procurable throughout Europe, except in very remote or unfrequented places.[1] In the

---

[1] Throughout France, St. Galmier; in Germany,

latter circumstances no admixture of wine or spirit counteracts the poison in tainted water and makes it safe to drink, as people often delight to believe; but the simple process of boiling it renders it perfectly harmless; and this result is readily attained in any locality by making weak tea, to be taken hot or cold; or in making toast-water, barley-water, lemonade, etc. The table waters now so largely imported into this country from Germany and France contain a considerable proportion of mineral matter in solution, and while they are wholesome as regards freedom from organic impurities, are, of course, less perfect for daily use than absolutely pure waters, such as those above referred to. Vaunted frequently as possessing certain medicinal properties, this very fact ought to prohibit their constant use as dietetic agents for

---

Selters; in Austria and Bohemia, Giesbübel, are always obtainable, being the table-water of most repute, in each case respectively, of the country itself. In all chief places in Italy, either Selters or St. Galmier, often both, are supplied by the hotels. In Spain these are not at present to be had, but the alternatives recommended are easily obtained.

habitual consumption, inasmuch as we do not require drugs as diet, but only as occasional correctives. Among them, the natural Selters, Apollinaris, Gieshübel, and St. Galmier — but of this latter some of the sources are inferior to others, the best appearing now to be chiefly retained for Paris — are perhaps among the most satisfactory within our reach. A dash of lemon-juice, and a thin cutting of the peel, form sometimes an agreeable addition. I am compelled to say that the sweet compounds and fruity juices which have of late been produced as dinner drinks, and apparently in competition with wine, are rarely wholesome adjuncts to a dinner. Such liquids rapidly develop indigestible acid products in the stomachs of many persons; while for all, the sipping of sweet fluids during a meal tends to diminish appetite, as well as the faculty of appreciating good cookery. If wine is refused, let the drink be of pure water — with a sparkle of gas in it, or a slight acid in it, if you will—but in obedience both to gastronomic and dietetic laws let it be free from sugar. No doubt there are exceptional circumstances in which fruity juices, if not very sweet, can be taken freely. Thus I have rarely quaffed more delicious liquor at dinner in the warm autumn of Southern Europe, notably in Spain, than that afforded by ample slices of a water-melon, which fill the mouth with cool fragrant liquid; so slight is the amount of solid matter, that it only just serves to contain the abundant delicate juices of the fruit grown in those climates. Here the saccharine matter is present only in small proportion.

Before concluding, a remark or two may be permitted in reference to that great British institution, the public dinner. Its utility must, I suppose, be conceded, since for a vast number of charitable and other interests the condition of commanding once a year the ear of the British public for an exposition of their claims, seems in no other way at present attainable. A royal or noble chairman, a portentous menu, an unstinted supply of wine, such as it is, and after-dinner speeches in variety, form an *ensemble* which appears to be attractive to the great body of " supporters." On the other hand, those whose presence is enforced by the claim of duty find these banquets too numerous and too long. The noise and bustle, the badly served although pretentious dinner, the glare of gas and the polluted air, the long, desultory, and unmeaning speeches, interspersed with musical performances—which, however admirable in themselves, extend unduly a programme already too comprehensive—unfit many a man, seriously occupied, for the engagements of the morrow. Might it not be worth trying the experiment of offering fewer dishes, better service, and abolishing half the toasts? Might it not be possible to limit the necessary and essential toasts of a public dinner to the number of three or four—these to be followed only by a few subordinate toasts associated with the minor interests of the special object of the dinner? With the utmost deference to long-received usage, and after some little consideration, I venture to suggest that the following programme would at all events be an improvement on the present system, if such it can be called:

The first toast, or toasts, by which we declare our fidelity to the Crown, and our loyalty to the person of the Sovereign, as well as to the Royal family, to remain, by universal consent, as before. The next, or patriotic toasts, unlike the preceding, are regarded as demanding response, often from several persons, and here it is that time is generally wasted. These might, therefore, be advantageously compressed into one, which need not be limited to the military and naval services, although it would of course include them. The object might be attained by constituting a single comprehensive but truly patriotic toast, viz., " Our great National Institutions," which are easily defined. Supposing them to be regarded as seven in number, a response might be provided for from any two, according to the speakers present and the nature of the special object. These institutions fall naturally into order, as — (1) Parliament : its leaders. (2) Justice : the judges. (3) The military and naval forces : their officers. (4) Education : heads of universities and public schools. (5) Religion : its ministers.

(6) Science and Art: heads of societies, academies, colleges. (7) Literature and the Press: distinguished writers.

The next to be "the toast of the evening:" in other words, the particular subject of the dinner. After this would follow the healths of officers connected with the subject, visitors, etc., if necessary.

I confess I see no reason why the military and naval forces, however profound our respect and our gratitude for their great services to the nation must be—and in this matter I yield to no man—should invariably occupy a toast and speech, to be responded to by at least two, often by three officers, while the other great and scarcely less important interests should be left out of consideration altogether, or be only occasionally introduced. The toast of "National Institutions" would mostly insure to the chairman and managers of the dinner an opportunity of obtaining two good speakers from different interests in reply—say, one for Justice, and the other for Religion; one for Parliament or the Services, and the other for Science or Literature, and so forth.

Thus all the varied elements of our national life would receive in their turn a due share of attention from the great mass of public diners, and better speeches would probably be secured than by the present mode.

I confess this is rather an episode; but the subject of "toasts" is so interwoven with the management of the public dinner, that I have ventured to introduce it. I even dare to think that the proposition may be not unlikely to receive the support of "the chair," the duties of which, with a long array of toasts, are sometimes excessively onerous; only more so, be it recollected, in degree than those, of a humbler kind, which are entailed on many of the guests who are compelled to assist.

In concluding this imperfect sketch of the very large subject indicated by the title of my paper, I desire to express my strong sense of its manifold shortcomings, especially by way of omission. Desiring to call attention, in the shortest possible compass, to a great number of what appear to me to be important considerations in connection with the arts of selecting, preparing, and serving food, I have doubtless often failed to be explicit in the effort to be brief. It would have been an easier task to illustrate these considerations at greater length, and to have exceeded the limits of a couple of articles; and I might thus perhaps also have avoided, in dealing with some topics, a tone in statement more positive than circumstances may have warranted. Gastronomic tastes necessarily differ, as races, habits, digestive force, and supplies of food also differ; and it becomes no man to be too dogmatic in treating of these matters. *De gustibus non est disputandum* is in no instance more true than in relation to the tastes of the palate. Still, if any rational canons are to be laid down in connection with food and feeding, it is absolutely necessary that something more than the chemical and physiological bearings of the subject should be taken into consideration. With these it is unquestionably essential for any one who treats of my subject to be familiar; but no less necessary is it to possess some natural taste and experience in the cultivation of the gustatory sense; just as a cultivation of the perception of color, and a sensibility to the charm of harmoniously combined tints, are necessary to an intelligent enjoyment of the visual sense, and to the understanding of its powers. Hence the treatment of the whole subject must inevitably be pervaded to some extent by the personal idiosyncrasy and predilections of the individual. It is this fact, no doubt, which, operating in relation to the numerous writers on cookery, has tended to produce some of the complication and confusion which often appears in culinary directions and receipts. But the gastronomic art is a simpler one than the effusions of some of its professors might lead the wholly uneducated to believe; and the complicated productions originated by some of its past and greatest practitioners are as unnecessary as are the long and complicated prescriptions formerly in vogue with the leading physicians of past time. Both were the natural outgrowths of an age when every branch of technical education was a "mystery;" and when those who had attained the meaning thereof magnified their craft in the eyes of the vulgar by obscuring what is simple in a cloud of pedantic terms and processes. But that age and its delusions are passing away, and it is high time for simplicity in the practice of cookery to take the place of some useless and extravagant combinations and treatment which tradition has handed down.

At the present day it appears desirable, before all things, to secure the highest quality of all produce, both animal and vegetable; a respectable standard being rarely attained throughout our country in regard to the products of the latter kingdom. Great Britain has long held, and still maintains, the first place as to quality for her beef and mutton; in no other country in Europe—I cannot speak of America —is it possible to obtain these meats so tender, juicy, and well developed. The saddle, the haunch, the sirloin, and the round, so admirable on occasions, are only in danger of suffering here, like intimate friends, from too great familiarity with their charms. But even our standard of quality in meat has been gradually lowered, from the closer struggle, year by year, to produce a fat animal in a shorter space of time than formerly; a result which is accomplished by commencing to feed almost exclusively on oil-cake at a very early period of life. The result of this process is, that size and weight are attained by a deposit of fat, rather than by the construction of muscular fibre, which alone is true meat; while, as a necessary consequence, the characteristic flavor and other qualities of fully developed beef and mutton are greatly wanting in modern meat.

Much more unsatisfactory is the supply of vegetable and dairy produce to our great city, particularly of the former. It must be confessed that our market at Covent Garden, in relation to capabilities for effective distribution of fresh vegetables, etc., would disgrace a town one-fifth of the size of London. Nineteen-twentieths of its inhabitants cannot obtain fresh green food on any terms, and those who succeed pay an exorbitant price. I think I am right in saying that a really new-laid egg is a luxury which a millionnaire can scarcely insure by purchase; he may keep fowls, and with due care may obtain it, not otherwise. The great staple of our bread, commonly called "baker's bread," is unpalatable and indigestible; and I suppose no thoughtful or prudent consumer would, unless compelled, eat it habitually—used as it nevertheless is by the great majority of the inhabitants of this great city—any more than he would select a steak from the coarse beef whose proper destination is the stock-pot. Let any one compare the facilities which exist in most foreign towns for obtaining the three important articles of diet just named, with the parallel conditions afforded by London, and the inferiority of the latter will be so manifest as to become matter of humiliation to an Englishman. I do not raise any question of comparison between our own markets and the Halles Centrales of Paris, covering as they do nearly five acres of closely utilized space, with enormous vaults beneath, in direct communication by tram-road with the railways; nor of the well-stocked Marché St. Honoré, and others of less note. To many among the thousands of tourists who frequent the public buildings of Paris, an early morning survey of the fish, flesh, dairy produce, vegetables, fruit, and flowers, which the Halles Centrales display, and the scarcely less remarkable exhibition of Parisian and provincial life brought together there, present one of the most interesting and truly foreign spectacles which the city affords.

To the long list of needed reforms I have ventured to advocate in connection with this subject, I must add the want of ample and accessible markets in various parts of London, for what is known as country produce. I do this not only in the interest of the millions who, like myself, are compelled to seek their food within the limits of Cockayne; but also in the interest of our country gardeners and housewives, who ought to be able to supply us with poultry, vegetables, and eggs, better than the gardeners and housewives of France, on whom at present we so largely depend. We may well be grateful to these small cultivators, who by their industry and energy supply our deficiencies; but the fact that they do so does not redound to the credit of our countrymen.

No doubt, as regards security, liberty, locomotive facilities, etc., Cockayne is a tolerably comfortable and pleasant place to live in; nevertheless it is certainly true that greater intelligence, more enterprise, and better organization—perhaps of the co-operative kind—are much required, in order to improve not only the sources and quality of our food, but also some of our manners and customs in relation to selecting, preparing, and serving it.

THE END.

A NEW LIBRARY EDITION
OF
Macaulay's England.

Macaulay's History of England. New Edition, from New Electrotype Plates. 5 vols., 8vo, Vellum Cloth with Paper Labels, Gilt Tops and Uncut Edges, $10 00. *Sold only in Sets.*

The beauty of the edition is the beauty of proper workmanship and solid worth, the beauty of fitness alone. Nowhere is the least effort made to decorate the volumes externally or internally. They are perfectly printed from new plates that have been made in the best manner, and with the most accurate understanding of what is needed; and they are solidly bound, with absolutely plain black cloth covers, without relief of any kind, except such as is afforded by the paper label. It is a set of plain, solid, sensible volumes, made for use, and so made as to be comfortable in the using.—*N. Y. Evening Post.*

Published by HARPER & BROTHERS, New York.

☞ *Sent by mail, postage prepaid, to any part of the United States, on receipt of the price.*

MOTLEY'S HISTORIES.

CHEAP EDITION.

THE RISE OF THE DUTCH REPUBLIC. A History. By JOHN LOTHROP MOTLEY, LL.D., D.C.L. With a Portrait of William of Orange. 3 volumes, 8vo, Vellum Cloth with Paper Labels, Uncut Edges and Gilt Tops, $6 00. *Sold only in Sets.*

HISTORY OF THE UNITED NETHERLANDS: from the Death of William the Silent to the Twelve-Years' Truce. With a full View of the English-Dutch Struggle against Spain, and of the Origin and Destruction of the Spanish Armada. By JOHN LOTHROP MOTLEY, LL.D., D.C.L. With Portraits. 4 vols., 8vo, Vellum Cloth with Paper Labels, Uncut Edges and Gilt Tops, $8 00. *Sold only in Sets.*

LIFE AND DEATH OF JOHN OF BARNEVELD, Advocate of Holland. With a View of the Primary Causes and Movements of the "Thirty-Years' War." By JOHN LOTHROP MOTLEY, LL.D., D.C.L. Illustrated. 2 vols., 8vo, Vellum Cloth with Paper Labels, Uncut Edges and Gilt Tops, $4 00. *Sold only in Sets.*

Published by HARPER & BROTHERS, New York.

☞ *Sent by mail, postage prepaid, to any part of the United States, on receipt of the price.*

OLIVER GOLDSMITH.

POETICAL WORKS OF OLIVER GOLDSMITH.
With a Biographical Memoir, and Notes on the Poems. Edited by BOLTON CORNEY. Illustrated. 8vo, Cloth, $3 00; Cloth, Gilt Edges, $3 75; Turkey Morocco, Gilt Edges, $7 50.

SELECT POEMS OF OLIVER GOLDSMITH.
Edited, with Notes, by WILLIAM J. ROLFE, A.M. Illustrated. Small 4to, Flexible Cloth, 70 cents; Paper, 50 cents.

GOLDSMITH'S POEMS.
32mo, Paper, 20 cents; Cloth, 35 cents.

GOLDSMITH'S PLAYS.
32mo, Paper, 25 cents; Cloth, 40 cents.

THE VICAR OF WAKEFIELD.
By OLIVER GOLDSMITH. 18mo, Cloth, 50 cents. 32mo, Paper, 25 cents; Cloth, 40 cents.

GOLDSMITH. By WILLIAM BLACK.
A Critical and Biographical Sketch. (In the series entitled "English Men of Letters.") 12mo, Cloth, 75 cents.

GOLDSMITH.—BUNYAN.—MADAME D'ARBLAY.
By Lord MACAULAY. 32mo, Paper, 25 cents; Cloth, 40 cents.

HISTORY OF GREECE.
By OLIVER GOLDSMITH. Abridged. 18mo, Cloth, 75 cents.

HISTORY OF ROME.
By OLIVER GOLDSMITH. Abridged. 18mo, Cloth, 75 cents.

OLIVER GOLDSMITH. By WASHINGTON IRVING.
With Selections from his Writings. 2 vols., 18mo, Cloth, $1 50.

Published by HARPER & BROTHERS, New York.

☞ *Any of the above works will be sent by mail, postage prepaid, to any part of the United States, on receipt of the price.*

SAMUEL JOHNSON.

BOSWELL'S JOHNSON. The Life of Samuel Johnson, LL.D. Including a Journal of a Tour to the Hebrides. By JAMES BOSWELL, Esq. Portrait of Boswell. 2 vols., 8vo, Cloth, $4 00; Sheep, $5 00; Half Calf, $8 50.

JOHNSON'S WORKS. The Complete Works of Samuel Johnson, LL.D. With an Essay on his Life and Genius, by ARTHUR MURPHY, Esq. 2 vols., 8vo, Cloth, $4 00; Sheep, $5 00; Half Calf, $8 50.

SKETCH OF JOHNSON. Samuel Johnson. By LESLIE STEPHEN. 12mo, Cloth, 75 cents.

JOHNSON'S LIFE AND WRITINGS. Selected and Arranged by the Rev. WILLIAM P. PAGE. 2 vols., 18mo, Cloth, $1 50.

MACAULAY'S JOHNSON. Samuel Johnson, L.L.D. By Lord MACAULAY. 32mo, Paper, 25 cents.

JOHNSON'S RELIGIOUS LIFE. The Religious Life and Death of Dr. Johnson. 12mo, Cloth, $1 50.

SAMUEL JOHNSON: His Words and his Ways; What he Said, What he Did, and What Men Thought and Spoke Concerning Him. Edited by E. T. MASON. 12mo, Cloth, $1 50.

PUBLISHED BY HARPER & BROTHERS, NEW YORK.

☞ HARPER & BROTHERS *will send any of the above works by mail, postage prepaid, to any part of the United States, on receipt of the price.*

☞ HARPER'S CATALOGUE *mailed free on receipt of Nine Cents in stamps.*

Holly's Modern Dwellings.

Modern Dwellings in Town and Country, adapted to American Wants and Climate. In a Series of One Hundred Original Designs, comprising Cottages, Villas, and Mansions. With a Treatise on Furniture and Decoration. By H. HUDSON HOLLY. 8vo Cloth, $4 00.

No feature of a house, whether important or insignificant, fails to receive its due share of attention. * * * The practical, sensible directions of Mr. Holly's book are numberless, and give it a character unlike many works, whose suggestions argue an inordinately plethoric purse and limitless cultivation in art. He is a business man, and deals in plans, specifications, and estimates. His volume contains many original designs of cottages, mansions, and villas, accompanied by complete descriptions in which the material, dimensions, and cost of building are distinctly stated in figures. * * * There is little doubt that the volume will prove, as its author suggests in his preface, "a practical and reliable guide for those persons who wish to build, furnish, and beautify their houses without an extravagant outlay of money."—*Boston Transcript.*

Mr. Holly's designs are not only tasteful, but his plans are arranged with an eye to comfort, plenty of closet room, and a convenient arrangement of rooms. * * * The hints are generally good in household art, and any one about to build will find it worth his while to consult Mr. Holly's book.—*N. Y. Herald.*

Published by HARPER & BROTHERS, New York.

☞ *Sent by mail, postage prepaid, on receipt of price.*

POTTERY AND PORCELAIN
OF ALL TIMES AND NATIONS.

With Tables of Factory and Artists' Marks, for the Use of Collectors.

By WILLIAM C. PRIME, LL.D.

ILLUSTRATED.

8vo, Cloth, Uncut Edges and Gilt Tops, $7. (In a Box.)

The outgrowth of the author's studies for years—not at all exclusive, but as an indulged and engrossing pastime—it gathers up in systematic form the results of his study, travel, and collections, but at the same time expands without hesitation upon any topic in his way that is favorite or more than usually familiar. Neglecting nothing in the proper scope of the work, it spares the reader much dry epitomizing from other works, gives the old information in a fresh way, and the author's special results with the ease and thoroughness of a master. * * * The book is a fit ornament for any library; but to ardent lovers of the science is simply a necessity.—*Sunday School Times,* Philadelphia.

The illustrations of the work are numerous and finely executed. The paper, printing, and binding are themselves a work of art, and sustain the established reputation of the publishers. It would be difficult to find a handsomer ornament than Dr. Prime's book for parlor or library table, or a more appropriate gift.—*N. Y. Times.*

Published by **HARPER & BROTHERS**, New York.

HARPER'S
Household Dickens Complete.

Elegant and Cheap. 8vo. With Original Illustrations by Thos. Nast, J. Barnard, E. A. Abbey, A. B. Frost, J. Mahoney, C. S. Reinhart, and other eminent American and English Artists.

Oliver Twist............ ⎫
A Tale of Two Cities... ⎬ Cloth, $1 00; Paper, 50 cents.
The Old Curiosity Shop...Cloth, $1 25; Paper, 75 cents.
David Copperfield...... ⎫
Dombey and Son...... |
Nicholas Nickleby..... |
Bleak House.......... |
Pickwick Papers....... |
Martin Chuzzlewit..... |
Little Dorrit.......... |
Barnaby Rudge ...... |
Our Mutual Friend..... ⎬ Cloth, $1 50; Paper, $1 00.
Christmas Stories...... |
Great Expectations.... |
The Uncommercial Traveller, Hard Times, and The Mystery of Edwin Drood............... |
Pictures from Italy, Sketches by Boz, and American Notes...... ⎭

The Set Complete, 16 vols., Cloth, in neat box, $22.

☞ HARPER & BROTHERS *will send either of the above works by mail, postage prepaid, to any part of the United States or Canada, on receipt of the price.*

BOOKS FOR YOUNG MEN.
By SAMUEL SMILES.

SELF-HELP: with Illustrations of Character, Conduct, and Perseverance. New Edition, revised and enlarged. 12mo, Cloth, $1 00.

> CONTENTS:—Spirit of Self-Help.—Leaders of Industry.—Three Great Potters.—Application and Perseverance.—Help and Opportunities.—Scientific Pursuits.—Workers in Art.—Industry and the Peerage.—Energy and Courage.—Men of Business.—Money, its Use and Abuse.—Self-Culture.—Facilities and Difficulties.—Example, Models.—The True Gentleman.

CHARACTER. 12mo, Cloth, $1 00.

> CONTENTS:—Influence of Character.—Home Power.—Companionship and Example.—Work.—Courage.—Self-Control.—Duty, Truthfulness.—Temper.—Manner, Art.—Companionship of Books.—Companionship in Marriage.—Discipline of Experience.

THRIFT. 12mo, Cloth, $1 00.

> CONTENTS:—Industry.—Habits of Thrift.—Improvidence.—Means of Saving.—Examples of Thrift.—Methods of Economy.—Life Assurance.—Savings-Banks.—Little Things.—Masters and Men.—The Crossleys.—Living above the Means.—Great Debtors.—Riches and Charity.—Healthy Homes.—Art of Living.

☞ HARPER & BROTHERS *will send either of the above works by mail, postage prepaid, to any part of the United States, on receipt of the price.*

www.ingramcontent.com/pod-product-compliance
Lightning Source LLC
Chambersburg PA
CBHW020111170426
43199CB00009B/490